T0336820

THE MATHEMATICAL THEORY OF NONBLOCKING SWITCHING NETWORKS

SERIES ON APPLIED MATHEMATICS

Editor-in-Chief: Zhong-ci Shi

*For the complete list of the volumes in this series, please visit
http://www.worldscientific.com/series/sam

Series on
Applied Mathematics
Volume 15

THE MATHEMATICAL THEORY OF NONBLOCKING SWITCHING NETWORKS

2nd Edition

Frank K Hwang

National Chiao-Tung University, Taiwan

World Scientific

NEW JERSEY · LONDON · SINGAPORE · BEIJING · SHANGHAI · HONG KONG · TAIPEI · CHENNAI

Published by

World Scientific Publishing Co. Pte. Ltd.

5 Toh Tuck Link, Singapore 596224

USA office: 27 Warren Street, Suite 401–402, Hackensack, NJ 07601

UK office: 57 Shelton Street, Covent Garden, London WC2H 9HE

British Library Cataloguing-in-Publication Data
A catalogue record for this book is available from the British Library.

THE MATHEMATICAL THEORY OF NONBLOCKING SWITCHING NETWORKS (2nd Edition)
Series on Applied Mathematics — Vol. 15

ISBN-13 978-981-256-042-1
ISBN-10 981-256-042-4

Printed in Singapore

Preface

The seed of this book had been planted during my twenty-eight years of employment at Bell Laboratories, where I learned everything about switching networks. The material of this book was first organized when I was teaching a one-semester graduate course with the same title at National Chiao-Tung University in 1997 Spring. I am grateful to the whole class for forcing me to work through the material, and especially to Prof. G. J. Chang, Prof. L. S. Hsu, Dr. H. G. Yeh, Dr. S. C. Liao and Dr. L. D. Tong who sat through the whole semester and helped me in clarifying many ideas. I thank Bell Laboratories for providing me a 3-month support during 1997 Summer to write a first draft, and for allowing Ms. Martina Sharp to do the excellent word-processing for the book, for which she sacrificed many nights and week-ends. The first four chapters were then taught at a special lecture series organized by the National Central University to test out the material, and the last chapter was given as a short course at the IEEE International Workshop on Broadband Switching.

Preface to the second edition

There are two main motives to publish a second edition. The first is to bring the discussions of the relevant topics up to date. In particular, there are many new results in the multicast and multirate chapters. These two chapters now constitute one half of the book instead of one third as in the first edition.

The second motif is to correct errors in the first edition, unfortunately, there are quite a few; and to improve the presentations in several places.

We hope that this new edition continues to serve the propose of keeping all relevant information on nonblocking switching networks in one place, and serve it better with an improved presentation of the material.

I would like to thank my three Ph.D students, B. C. Lin, J. Y. Guo and F. H. Chang for carefully proofreading through the book, especially, J. Y. Guo has helped in many editing task. Of course, the responsibility of any error or misrepresentation is completely mine.

Contents

Chapter 1. Introduction

1.1 Two-sided Nonblocking Switching Networks

The need of a switching network first came from the need to interconnect pairs of telephones. At first, when there were not that many phones, a direct wire was installed between every two phones. But with the increase in the number of phones, the transmission cost of these wires became overbearing and the notion of switching was born. Every phone in a given locality was then connected to a "switching" center where the wires from these phones were interconnected through a network called *switching networks*. Later, it was reinvented for the parallel computer to interconnect a set of processors with a set of memories. Currently, it is intended for many other applications, data transmission, video rental, conference calls, broadcast, satellite communication ... . It is safe to say that the need of switching network is expanding fast.

A switching network can either interconnect one group of users, called a 1-sided network, or two groups, called a 2-sided network. While there are applications for the one-sided network and a theory has been developed for it, the dominant applications and theory for switching networks are 2-sided. For many applications, the two sides represent two genuinely different types of entity; so input x connecting to output y is not the same as input y connecting to output x. Even for the telephone network which seems to interconnect only one group of users, the real networks could still be 2-sided because they might connect customers' wires to line equipments, or two different sets of customers, or two sets of time slots as in time division switching (here again, input i connecting to output j doesn't imply input j connecting to output i). Note that a 2-sided network can be used as a 1-sided network by putting the same set of entities on both sides, although this is less economical from the switching viewpoint. In this book, we will only deal with 2-sided networks.

In the 2-sided case we assume that the network has a set of input terminals and a set of output terminals, while the former generate requests to be connected to the latter through the network. Theoretically, an input terminal can request to be connected to any output terminal, just as one phone can call any

1

other phone. Therefore the network must provide access from any input terminal to every output terminal. Furthermore, once a connection is established,
it could last for a period of time, while other input terminals may generate
their own requests during this period. What a switching network does is to simultaneously connect these requests, the pattern constantly changing by some
terminals hanging up and others making new requests. For economic reason,
the network usually doesn't provide a dedicated path from an input terminal
to an output terminal, but provides a common pool of connection links. Thus
the connection of one request may block that of another. It was an amazing
achievement that Clos (1953) first showed that through some clever design
supported by some mathematical principle, there exist nonblocking networks
with significantly less hardware than a network with dedicated lines.

For telephone networks, perhaps "nonblockingness" is too high a goal.
First of all, not all customers are calling simultaneously, so even a blocking
network can be satisfactory for most practical purposes. Secondly, when a call
is blocked, the customer usually accepts the situation with grace and makes
a second attempt at a later time. It is only frequent blocking that may lead
to a complaint, and perhaps, a loss of revenue. However, many other applications have a lower tolerance for blockage, and the cost of blockage can be
higher. For example, when a customer moves to a nearby place and requests
the same phone number, the network interconnecting the customer's wires with
line equipments (phone numbers) is not expected to reject the request due to
blockage. Military or high-security networks are usually nonblocking. Blockage in a data-network can lead to a serious problem if the data happens to
be important. Customers of a video-rental network would not wait beyond
the starting time of the program. On the other hand, the cost of hardware
has been dropping fast, which makes nonblocking networks more and more a
realistic alternative, and even a necessity sometimes.

While we recognize that most switching networks in the world are still the
blocking kind, and there is a very interesting theory developed for it, this book
will deal with nonblocking, or almost nonblocking networks only because the
theory of nonblocking networks is very different from that of blocking networks.
By doing so, we will miss many important topics like packet switching, queue-
ing networks, buffers, traffic balancing, blocking probability, etc. However, we
think that setting up a framework for nonblocking networks to accommodate
their many existing beautiful results is itself a worthwhile task.

Not many books have been published on the theory of switching networks
since the Beneš (1965) classic. Varma–Raghavendra (1994) edited a collection

of benchmark papers with a detailed introduction before every topic. Hui's book (1990) covers not only switching, but also traffic theory. It does not focus on the mathematical theory as our book intends to. Li's book (2001) emphasizes on the engineering and application side of switching network, but touches only the most fundamental part of mathematical theory.

1.2 Networks

The basic components of a switching network are crossbar switches, or just *crossbars* (occasionally other switches are used and will be referred to as *switches*), and links which connect crossbars. A crossbar with n inlets and m outlets, denoted by X_{nm}, is said of size $n \times m$. Inlets (outlets) on the same crossbar are called *co-inlets* (*co-outlets*). Any matching (one-to-one mapping) between the inlets and the outlets of a crossbar is considered routable, i.e., a crossbar is nonblocking. Some crossbars are connected to the outside world. For a 2-sided network, one set of such crossbars will be called *input crossbars* and another set *output crossbars*. The links on an input (output) crossbar linking to outside world are called *inputs* (*outputs*) of the network, and often drawn by open-ended lines. They are also referred to as *external links*, while other links are *internal links*. An (N, M)-network has N inputs and M outputs, which will be called an *N-network* if $M = N$. Although a request is originally generated by a pair of input-output, it can be treated as if generated by a pair of input-output crossbars since the crossbar is nonblocking. A request is connected by a path in the network, while two connections do not block each other if their paths are link-disjoint.

We shall warn the reader that when the switching network is intended to be used as a computer network such as a processor-memory network, then it is customary to assume that the input and output crossbars have no external links. So the numbers of inputs and outputs are simply the numbers of input and output crossbars, respectively. We would not attempt to use different names to differentiate networks with or without external links, since the mathematical theory for them is usually the same. We will call the reader's attention whenever a difference is relevant, and draw the input and output crossbars in circles when no external links are intended, as versus the usual squares for crossbars.

In an s-stage network, the crossbars are lined up into s columns, each called a *stage*. Sometimes s is not specified and the network is called a *multistage interconnection network* (MIN). Crossbars in the same stage have the same

size. Links exist only between crossbars in adjacent stages. Links between a
stage-i crossbar and a stage-$(i+1)$ crossbar connects an outlet of the former to
an inlet of the latter. Often, only the information of which stage-i crossbars are
connected to which stage-$(i+1)$ crossbars are needed. Then the linking pattern
between stage i and stage $i+1$ can be represented by a bipartite graph L_i
with the crossbars as vertices. Crossbars in the first (last) stage are the input
(output) crossbars and its inlets (outlets) are the input (output) terminals,
sometimes just called inputs (outputs) of the network connected to external
lines. The notation for an s-stage network is that stage i has r_i crossbars of
size $n_i \times m_i$. Necessarily, $r_i m_i = r_{i+1} n_{i+1}$ for $i = 1, \ldots, s - 1$. Figure 1.2.1
shows a 3-stage network with 8 inputs and 6 outputs where $r_1 = r_2 = 4$,
$r_3 = 2$, $n_1 = 2$, $m_1 = 3$, $n_2 = 3$, $m_2 = 1$, $n_3 = 2$, $m_3 = 3$, and a crossbar is
represented by a square.

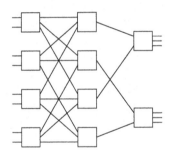

Figure 1.2.1: A 3-stage network

Two networks are called equivalent if one can be obtained from the other
through permuting crossbars in the same stage.

Let $X_{ab} \times X_{cd}$ denote the 2-stage network where stage-1 consists of c copies
of X_{ab}, stage-2 consists of b copies of X_{cd}, and the linking pattern between the
two stages is a complete bipartite graph. Let $[X_{ab}, X_{cd}, X_{be}]$ denote the 3-
stage network where stage-1 consists of c copies of X_{ab}, stage-2 consists of
b copies of X_{cd}, stage-3 consists of d copies of X_{be}, and the linking patterns
between adjacent stages are complete bipartite graphs. These notations, due
to Cantor (1971), are often used to easily describe the structure of a MIN.
Figure 1.2.2 illustrates these networks. The 3-stage network $[X_{ab}, X_{cd}, X_{be}]$ was
first proposed by Clos (1953) and is now known as the *3-stage Clos network*.
The traditional notation for 3-stage Clos network is $[X_{n_1 m}, X_{r_1 r_2}, X_{m n_2}]$, where
$N_1 = n_1 r_1$ is the number of inputs and $N_2 = n_2 r_2$ is the number of outputs.

Note that the use of m, n_2, r_2, N_2 is not consistent with the notation for MIN.

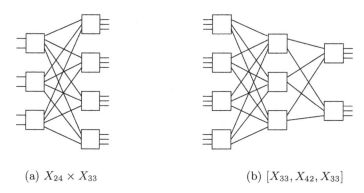

(a) $X_{24} \times X_{33}$ (b) $[X_{33}, X_{42}, X_{33}]$

Figure 1.2.2: Two constructions of networks

The notation introduced in the last paragraph can be extended by replacing the crossbars by networks. For example, the well known $(2n-1)$-stage *Beneš network* B_n is defined recursively by $B_2 = [X_{22}, X_{22}, X_{22}]$ and $B_n = [X_{22}, B_{n-1}, X_{22}]$, where the number of the first(last) X_{22} equals the number of inputs(outputs) in B_{n-1}, which is 2^{n-1}. Another network operator often used is *concatenation*. Let ν denote an s-stage network and ν' an s'-stage network such that the last stage of ν is identical to the first stage of ν'. Then $\nu \circ \nu'$ denote the $(s + s' - 1)$-stage network obtained by identifying the last stage of ν with the first stage of ν'. Figure 1.2.3 illustrates the concatenation of the two networks in Fig. 1.2.2.

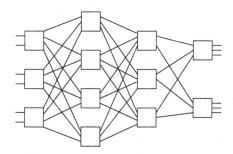

Figure 1.2.3: Concatenation of the two networks in Fig. 1.2.2.

A *d-nary network* (MIN) is simply a network (a MIN) using only crossbars of size $d \times d$. In a d-nary MIN of size N, a power of d, it is customary to use the notation $n = \log_d N$(but if input and output switches have no external link, then $n = \log_d N + 1$), which we adopt in this book. Note that in a d-nary MIN every stage has the same number of crossbars. The term *almost d-nary* is used if exceptions are allowed for the input and output stages. The *d-nary Beneš network*, denoted by B_n^d, is similarly defined as the Beneš network except using $d \times d$ crossbars.

Figure 1.2.4 illustrates two d-nary MINs.

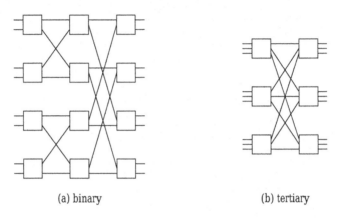

(a) binary   (b) tertiary

Figure 1.2.4: Two d-nary MINs

The notion of a *staged* network was first proposed by Halpenny (1990). The only requirement is that the network can be laid out in the plane with input terminals on one side, output terminals on another, and a link, represented by a straight line, does not bypass a crossbar. A network is *planar* if it can be laid out in the plane such that links do not cross each other. Figure 1.2.5 illustrates these networks. The relevance of the planar network to the optical switch was first called into attention by Spanke and Beneš (1987).

1.3   Traffic and Nonblockingness

When we say that a network is nonblocking, it is in reference to a certain type of traffic. We can classify traffic according to whether the requests come one by one, like phone calls, or they are scheduled into sessions, and all requests in a given session, called a *frame*, are given simultaneously for routing, like video-rental or satellite communication. We call the former type *dynamic* and the latter *scheduled*. For convenience, we also refer to all traffic currently carried

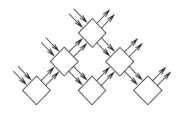 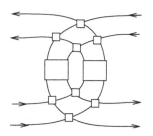

(a) A staged (also planar) network with six input (output) terminals

(b) A planar network with four input (output) terminals

Figure 1.2.5: A staged network and a planar network

by the network plus the new request in the dynamic traffic as a frame.

The traffic can also be classified as *point-to-point*, like 2-party phone calls, or *broadcast*, which is one to many. If there is a restriction on the maximum number of receivers per request, then broadcast is called *multicast*, or *f-cast*, if that number is specified to be f. The dynamic multicasting traffic can be further divided into two types according to whether additional receivers can be added after a multicast request is already connected. We will use *open-end traffic* (which allows additions) and *closed-end traffic* (which does not allow) to differentiate the two types.

Finally, we call the traffic *classic* if each link (terminal) carries (generates) one request and *multirate*, if each request is associated with a load, while each link (terminal) has a capacity and can carry (generate) as many requests as desirable as long as the total load does not exceed the capacity. Such kind of traffic can be generated by networks which integrate various types of requests like phone calls, data transmission, video signals with different bandwidth requirements (loads). If a network is designed for a special type of traffic, we can also transfer the classification of traffic to networks. Note that for the classic traffic, capacity = load = unity.

Define the *states* of a network as the set of all possible routings of all legitimate frames, *legitimate* means the load generated by each input and output terminal does not exceed its capacity. The empty state is simply the state of carrying no traffic. The states can be partially ordered by containment where *s containing s'* means s can be obtained from s' by adding the routings of more requests (but still legitimate). A set of requests is routable if there exists a set of link-disjoint paths connecting the requests. A state is *blocking* if there exists a legitimate new request not routable in the current state; and is *nonblocking* otherwise.

Traditionally, there are different levels of nonblockingness: strictly, wide-sense and rearrangeable. A network is *strictly nonblocking* (SNB) if it has no blocking state. Beneš (1965) proposed the concept of *wide-sense nonblocking* (WSNB). A network is WSNB under a routing algorithm A if there exists a closure CL(S) of states called *safe states*, closed under the operation of adding or deleting a connection, such that

(i) a safe state is a nonblocking state,

(ii) the empty state is a safe state,

(iii) any legitimate request generated in a safe state can be routed according to A into another safe state.

Sometimes A is given as a class. What it means is that the network is WSNB for any algorithm in A. Other times A is not mentioned, what it means is that there exists an algorithm under which the network is WSNB. With the same logic, a network is not WSNB means under no algorithm is it WSNB.

Smyth (1988) gave a method to find CL(S) under a routing algorithm A. First delete all blocking states. Then iteratively delete all states which may be forced to go to a deleted state under A by a deletion or an addition. The undeleted states, if any, constitute CL(S). If CL(S) is empty, then the network is not WSNB under A.

Since a routing is done by setting (the internal connection of) the switches, a state can also be defined as a setting of switches. The difference between SNB and WSNB is actually more subtle than it appears since what are the possible settings of a switch is often hidden. Take the crossbar X_{nm}, which is usually considered SNB and represented as a grid of n rows and m columns. At each row-column intersection, there is a crosspoint with two states "straight" and "bend", which controls the routing (see Fig. 1.3.1).

If each crosspoint is free to go straight or bend, then we claim that a crossbar is not SNB. Figure 1.3.2 shows an X_{33} where the $(2,2)$ path blocks the $(1,3)$ and the $(3,1)$ path.

Of course, by not allowing a path to make right turns, we can avoid the above situation and show nonblockingness. However, then a crossbar is merely WSNB, because there exist blocking states; it is only that we don't get into them by directing traffic cleverly. On the other hand we can wire a crosspoint

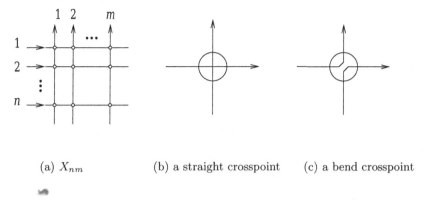

(a) X_{nm} (b) a straight crosspoint (c) a bend crosspoint

Figure 1.3.1: Crossbar and crosspoint

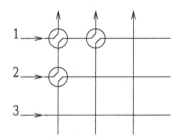

Figure 1.3.2: X_{33} is not SNB

differently to disallow the right turn (see Fig. 1.3.3), then a crossbar remains SNB.

Thus we see that the difference between SNB and WSNB hinges on more details about the hardware, which is unfortunate because we now have to provide the details of hardware, for example, which type of crosspoints, before discussing the nonblocking property of a network. (This is especially awkward for a mathematician who usually doesn't know the engineering details.) Of course, if we know a component can be implemented to be SNB, then instead of giving the engineering details, we can simply specify a SNB component. Another alternative, which seems to be the common practice as far as the crossbar is concerned, is to assume a component is SNB if it can be implemented that way.

Perrier–Prucnal (1989) proposed a notion of nonblockingness for point-

Figure 1.3.3: A differently wired crosspoints

to-point traffic which also blurs the difference between SNB and WSNB. A network is a *standard path network* (also called *fixed routing*) if it provides a standard path for each pair of input-output terminals to route their request, and the standard paths in any legitimate frame are all disjoint. Note that this doesn't mean that the standard paths are all disjoint, since it is possible that the standard paths of $(1,3)$ and $(2,3)$ overlap, but these two requests cannot coexist in a legitimate frame. A standard path network is clearly WSNB where the set of nonblocking states are those states which use standard paths. On the other hand, a standard path network can often be hard-wired into a SNB network by physically handicapping all paths except the standard ones. For example, the crossbar as shown in Fig. 1.3.1 is a standard path network where the standard path for the (i,j) request is to turn at the (i,j) intersection. This standard path crossbar can be turned into SNB by hard-wiring the crosspoints as shown in Fig. 1.3.3. Note that a general WSNB network cannot be hard-wired into SNB because the connection paths, unlike the standard paths, are traffic dependent.

For scheduled traffic, a network which can route all legitimate frames is called rearrangeably nonblocking(RNB), or simply *rearrangeable*. Such a network can also route dynamic traffic by rerouting some existing connections to make room for a newly blocked request, hence the name "rearrangeable". Note that by rerouting all existing calls, dynamic traffic is turned into scheduled traffic. A rearrangeable network with point-to-point traffic is often called a *connector*. We will also call one with multicast traffic a *multicast connector*, which sounds more specific than the term *generalizer* used in the literature.

A related notion first proposed by Ackroyd (1979) is the *repackable* network. For dynamic traffic, the rerouting is done not when a new request is blocked, but when an existing connection is deleted, with the purpose to balance the load carried by the intermediate switches. The rerouting is usually limited to one connection. A network is *repackable* if it is nonblocking by following the rerouting rule. Conceptually, repackability is also similar to WSNB except the traffic being directed is the existing connections rather than the new arrivals.

1.4 Objectives

Traditionally, nonblocking networks are designed to minimize the number of crosspoints since that is the most expensive part in a network. Even though in many new technologies, the cost of crosspoints is no longer a dominant issue, the number of crosspoints still remains a popular measure of network performance since it serves a figure of merit for some other important features, like the physical size and control complexity of a network, or the number of ports required when a network is partitioned into chips or boards. We define the *cost* of a network as the number of crosspoints in it.

In a multistage network, the number of stages represents the length of a path. A shorter path of course means a faster connection, less likely to encounter a faulty element or to have the signal being transmitted weakened, and to consume less power. Under the current technology, only networks with a very small number of stages are practical.

Routing algorithms tend to be overlooked in SNB networks since a free path is guaranteed for any request even no routing algorithm is given. However, an efficient routing algorithm still serves to speed up the connections. Routing algorithms of course play a more fundamental role in WSNB and rearrangeable networks since the very claim of nonblockingness depends on the existence of a routing algorithm. Such routing algorithms are evaluated by their time complexity. Any algorithm requiring more than linear time would be too complicated for real-time use. Two remedies are to use parallel processors to route requests and to use self-routing algorithms, which provide standard paths in a blocking network. The trade-off is a higher cost for the former remedy, and giving up nonblocking in the latter.

To partition a switching network and to lay it out into boards or chips, the criteria are the modularity of the network, the number of parts, the number of ports on a part, the number of knock-knees and crossings, and planarity. For optical switches, the crosstalk problem is of particular importance and the planar network is one solution to that problem. Research on these problems are emerging, but will not be covered in this book (except planar networks).

The literature on switching networks shows two different lines of approach in constructing nonblocking networks with the number of crosspoints as objective. One is to consider networks of practical sizes and to analyze their performances. The other is a theoretical pursue of networks with the minimum orders of complexity. While the latter approach usually produces better theoretical results, one should be warned that some of the complexity results are proved by existence without explicit constructions, some of the constructions may have very large coefficients which are not shown in the orders of magnitude complexity, some of the results can be achieved only with a very large number of stages, or an unbounded number of terminals. For these reasons, we will focus on the practical networks, but provide the theoretical results as a background. In particular, some constructions rely heavily on the constructions of bipartite graphs with certain properties. While these graph constructions are very clever and interesting of their own right, we choose not to delve into them since they are more graph-theoretic than network-theoretic. If we need them in a network, we simply quote their properties from known sources.

1.5   Self-routing Networks and Their Extra-stage Version

For computer networks, delays more than polylog time are generally unacceptable. Therefore centralized routing algorithms which usually require $O(N \log N)$ time are out. Instead, a bunch of $\log_2 N$-stage networks with self-routing property have been invented; here, *self-routing*, first proposed by Lawrie (1976) for the Omega network, means that a request can be routed by only knowing its input and output, and nothing about other requests. These networks are usually recognized as the banyan-type by the following features.

(i) The network is an n-stage binary network $(n = \log_2 N)$.

(ii) Each input has a unique path to each output.

Agrawal (1983) called attention to another common property of there self-routing networks. Let u and v be two stage-k crossbars, and let $V_j(u)$ and $V_j(v)$ denote the two sets of stage-$(k + j)$ crossbars u and v can reach. Then the network is said to have the (k,j) *buddy property* if either $V_j(u) \cap V_j(v) = \emptyset$ or $V_j(u) = V_j(v)$ for all u, v. If a network has the (k,j) buddy property for all k and j, it is a *buddy network*.

Some well-known self-routing networks which have the buddy property, are shown in Fig. 1.5.1.

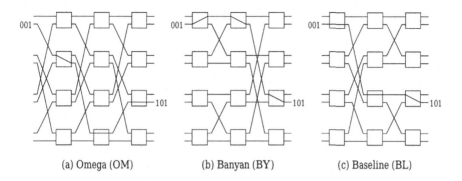

Figure 1.5.1: Some self-routing networks

The above class of binary networks can be extended to d-nary by replacing (i) with (i') $N = M = d^n$. The network consists of n stages of X_{dd}'s.

Note that an s-stage d-nary network is completely determined by the linking patterns between stages. Two such networks are called *equivalent* if the crossbars can be labeled such that the linking functions of the two networks become identical. Parker (1980) established the equivalence of some n-stage binary networks. Wu–Feng (1980) systematically studied a set of popular n-stage binary networks and proved them to be all equivalent. Agrawal (1983) proposed that the buddy property characterizes this equivalence class, but Bermond–Fourneau–Jean-Marie (1987) gave a counter-example. Instead, they gave the following characterization.

A binary n-stage network is said to have the $P(i,j)$ property if the sub-network from stage i to stage j, viewed as a graph, has exactly $2^{n-1-(j-i)}$ components.

A binary n-stage network is said to have the $P(\cdot,\cdot)$ property if it has the $P(i,j)$ property for all $1 \leq i \leq j \leq n$.

Theorem 1.5.1. *A banyan-type network is equivalent to the Omega network if and only if it has the $P(\cdot,\cdot)$ property.*

The proof of Theorem 1.5.1 is withheld as we are going to prove a more general result, which does not restrict the network to n stages.

Kruskal-Snir (1986) defined the *bidelta* network BD_n as one which can be recursively constructed in both directions, i.e., $BD_1 = X_{dd}$, $BD_n = X_{dd} \times BD_{n-1} = BD_{n-1} \times X_{dd}$. They independently proved the following result which can now be stated as a corollary of the d-nary version of Theorem 1.5.1.

Corollary 1.5.2. *All n-stage bidelta networks are equivalent.*

Let W^{-1} denote the inverse network of W, i.e., reversing the order of the stages. An $(n+1)$-stage buddy network was first proposed by Siegal–Smith (1978) for increasing the connection power and for fault tolerance. Shyy–Lea (1991) considered adding k extra stages to BY^{-1} and specified that the extra k stages should be identical to the mirror image of the first k stages. Represent a k-extra-stage buddy network by $B(k,n)$ or $B(k,N)$. The specified way of addition has the advantage that $BY^{-1}(k,n)$ can be sequentially decomposed j times, $1 \le j \le k$, namely the subnetwork of $BY^{-1}(k,n)$ from stage $j+1$ to stage $n+k-j$ decomposed into 2^j $BY^{-1}(k-j,n-j)$ such that each input (output) switch of the $BY^{-1}(k,n)$ has a unique path to each $BY^{-1}(k-j,n-j)$ (see Fig. 1.5.2 in which the external terminals are not drawn). Denote this way of adding extra stages by F^{-1}. Hwang–Liaw–Yeh (1998) observed that there are three other natural ways of addition.

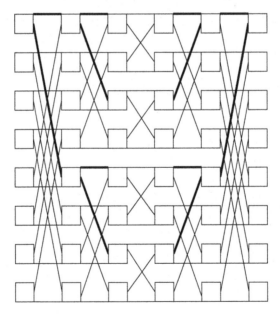

Figure 1.5.2: Decomposition of $BY^{-1}(2,4)$

(i) F: The extra k stages are identical to the first k stages.

(ii) L: The extra k stages are identical to the last k stages.

(iii) L^{-1}: The extra k stages are identical to the mirror image of the last k stages.

The various ways of addition result in different networks with different connection capabilities in general. Extra-stage/Omega networks are known as *shuffle exchange* (SE) networks. Hwang–Liaw–Yeh determined the equivalence classes among the k-extra-stage networks $SE(k)$, $SE^{-1}(k)$, $BY(k)$, $BY^{-1}(k)$, $BL(k)$, $BL^{-1}(k)$ for all k and under each of F, F^{-1}, L, L^{-1}.

Chang, Hwang and Tong (1999) proposed the class of bit permutation networks. Label the crossbars in a stage by distinct d-nary $(n-1)$-sequence. A bit-i group (or simply i-group) consists of the d crossbars whose labels differ only in bit i (there are d^{n-2} bit-i groups for each i). Let G_i denote the bipartite graph connecting the crossbars of stage i and stage $i+1$. An s-stage network is a bit permutation network if every G_i, $1 \leq i \leq s-1$, corresponds to a mapping from bit-u_i groups of stage i to bit-v_{i+1} groups of stage $i+1$ for some u_i, v_{i+1}. They proved that a bit permutation network is equivalent to one whose G_i corresponds to a mapping with $v_{i+1} = u_i$ for all i. Such a network can be characterized by the vector $(u_1, \cdots, u_{s-1})$. Further, they showed that two characterizing vectors are equivalent if one can be obtained from the other through permuting the $n-1$ bits.

Recently, Li (2001) proposed to view G_i as a bipartite graph between the outputs of stage i and the inputs of stage $i+1$. Label the outputs (inputs) by distinct d-nary n-sequences. Then G_i gives a bijection from the d^n outputs to the d^n inputs, and hence can be treated as a permutation. Such a permutation is called a *bit permutation* if it can be characterized by a permutation Π_i of the n bits. We will call a network an *edge bit permutation network* if each G_i corresponds to a Π_i. Since a permutation Π mapping bit n to bit n will cause all outputs of a stage i crossbar connected to the same stage-$(i+1)$ crossbar (highly undesirable), we assume $\Pi(n) \neq n$ throughout this section.

Hwang (2003) established a framework to properly place the above-mentioned four classes of networks: buddy(B), $P(\cdot, \cdot)$, bit permutation (BP) and edge bit permutation (EBP). Since $P(\cdot, \cdot)$ is defined only for n-stage networks, he generalized it to the class of power-of-d networks (d^P). An s-stage network is in this class if for any i, j, $1 \leq i < j \leq s$, the number of components in G_{ij} is a power of d. The intersection of the power-of-d class and the buddy class is called the power-of-d buddy class ($d^P B$).

The permutation in Fig. 1.5.3 (the stages are drawn in horizontal to save space) illustrates a bit permutation (1432) where x_1, x_2, x_3, x_4 are mapped to x_2, x_3, x_4, x_1.

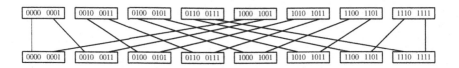

Figure 1.5.3: a bit permutation

We now show that a bit permutation T_i defines a mapping from u-groups of stage i to v-groups of stage $i+1$. In fact, we can pinpoint u and v.

Lemma 1.5.3. *Suppose G_i is represented by the bit permutation Π_i. Then G_i induces a mapping from $\Pi_i^{-1}(n)$-groups to $\Pi_i(n)$-groups.*

Proof. Note that the label of a crossbar can be obtained from the labels of its d edges by dropping the last bit. Since $\Pi^{-1}(n)$ is mapped to n and get dropped in the crossbar label of stage $i+1$, the d stage-i crossbars differing only in bit $\Pi^{-1}(n)$, i.e., the $\Pi^{-1}(n)$-group, are mapped to the same set of stage-$(i+1)$ crossbars.

On the other hand, the stage-i crossbar containing d edges whose left endpoints differ only in bit n is mapped to the $\Pi_i(n)$-group of stage $i+1$. Lemma 1.5.3 is proved. □

For the example in Fig. 1.5.3, the mapping is from $(\Pi_i^{-1}(4) = 1)$-groups to $(\Pi_i(4) = 3)$-groups.

Corollary 1.5.4. *$EBP \subset BP$.*

Proof. That $EBP \subseteq BP$ is trivial. The strict containment is shown by Fig. 1.5.4 which gives a network in BP (G_1 maps 3-groups to 3-groups, G_2 maps 2-groups to 3-groups), but not in EBP (Π_2 is not a bit permutation). □

Next we give a simple proof of vector-characterization of an EBP network.

Lemma 1.5.5. *Suppose G_i corresponds to a bit permutation Π_i which maps $\Pi_i^{-1}(n)$-groups of stage i to $\Pi_i(n)$-groups of stage $i+1$. Suppose we permute the crossbars of stage $i+1$ such that the jth crossbars of the $\Pi_i(n)$-groups are lined up with the jth crossbars of the $\Pi_i^{-1}(n)$-group, $j = 0, 1, \cdots, d-1$. Then the permutation Π (of stage $i+1$ crossbars) can be obtained from Π_i^{-1} by dropping n.*

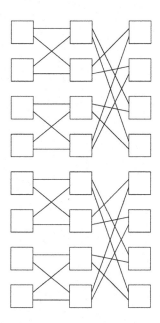

Figure 1.5.4: A *BP* network

Proof. Take a $\Pi_i^{-1}(n)$-group of stage i. Then the jth crossbars in this group is mapped (lined up) to the jth crossbars of the corresponding $\Pi(n)$-group of stage i to the crossbars of stage $i + 1$ under this lined-up operation. Then the only difference between Π_i and Π_i' is that in Π_i, $\Pi_i^{-1}(n)$ maps to n and n maps to $\Pi_i(n)$, while in Π_i', $\Pi_i^{-1}(n)$ maps to $\Pi_i(n)$. Therefore Π_i' can be obtained from Π_i by dropping n. But what we look for is the mapping $(\Pi_i')^{-1}$, namely, to move each stage $i + 1$ crossbar to the position its stage-i mate occupies. Hence Lemma 1.5.5.   □

In Fig. 1.5.5, $\Pi_i = (123)$, G_i is a bit permutation from 2-groups to 1-groups. The lining up of the stage-$(i+1)$ crossbars corresponds to a permutation $\Pi_i = (12)$ of stage-$(i+1)$ crossbars. Note that Π_i' can be obtained from $\Pi_i^{-1} = (132)$ by dropping 3.

Corollary 1.5.6. *A group in stage $i + 1$ remains a group after Π_i'.*

Proof. Since Π_i' merely permutes the $n-1$ bits of the crossbar labels, a u-group is transformed to a $\Pi_i'(n)$-group.   □

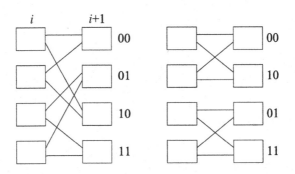

Figure 1.5.5: Lining up stage-$(i+1)$ crossbars

This leads to a proof of the vector characterization of the BP network simpler than the original proof by Chang, Hwang and Tong.

Theorem 1.5.7. *Consider an s-stage bit permutation network. By permuting the crossbars at stage 2, 3, $\cdots$, s, each G_i can be characterized by mapping u_i'-groups to u_i'-groups, $i = 1, \cdots, s-1$.*

Proof. We prove Theorem 1.5.7 induction on s. Theorem 1.5.7 is trivially true for $s = 2$ since we can permute the crossbars of stage 2 to line up with their mates in stage 1. Then G_1 becomes mapping u_1-groups to u_1-groups. Suppose Theorem 1.5.7 holds for up to $s-1$ stage. We prove for s stages.

Again, permute the crossbars at stage 2 according to $(\Pi_1')^{-1}$ to line up with their mates at stage 1. By Corollary 1.5.6, a u_2-group at stage 2 remains to be a group, say, a u_2'-group. Thus we may apply induction on the $(s-1)$-stage bit permutation network such that G_i is characterized by mapping u_i'-groups to u_i'-groups for $i = 1, 2, \cdots, s-1$.   □

Corollary 1.5.8.

$$u_i' = (\Pi_1')^{-1}(\Pi_2')^{-1} \cdots (\Pi_{i-1}')^{-1}(u_i)$$

For the example in Fig. 1.5.3, $\Pi_i = (1432)$ and the mapping is from 1-groups to 3-groups and $(\Pi_i')^{-1} = (123)$ for all $i = 1, 2, 3$. Thus $u_1' = u_1 = 1$, $u_2' = (\Pi_1')^{-1}(1) = 2$, $u_3' = (\Pi_1')^{-1}(\Pi_2')^{-1}(1) = (\Pi_1')^{-1}(2) = 3$.

Although we showed $BP \subset EBP$ in Corollary 1.5.4, the two classes are equivalent.

Theorem 1.5.9. *A BP network is equivalent to an EBP network.*

Proof. The vector characterization of a BP network actually means that a BP network is equivalent to another BP network whose mapping in G_i is bit-u_i group for $i = 1, \cdots, s - 1$. Since the mapping bit-u_i group to bit-u_i group corresponds to the edge bit-permutation (u_i, n), Theorem 1.5.9 follows. $\square$

Chang, Hwang and Tong proved

Theorem 1.5.10. *Suppose a d-nary bit permutation network is characterized by the vector $(u_1, \cdots, u_{s-1})$ which contains k distinct elements. Then the network has d^{n-1-k} components.*

Proof. Let i, $1 \le i \le n - 1$ be a number not in $(u_1, \cdots, u_{s-1})$. Then the crossbars on a path never change their i^{th} bit, i.e., all crossbars in a component have the same i^{th} bit. Since there are $n - 1 - k$ such i not in $(u_1, \cdots, u_{s-1})$, they generate d^{n-1-k} different combinations and hence that many components. $\square$

Corollary 1.5.11. BP $\subseteq d^P$.

Theorem 1.5.12. BP $\subseteq B$.

Proof. Consider an s-stage BP network characterized by $(u_1, \cdots, u_{s-1})$. Let v be a crossbar in stage i which reaches a set $V_j(v)$ of crossbars in stage j. Then $V_j(v)$ consists of crossbars whose labels are same as v'_s in bits in the set $I = \{1, \cdots, n-1\} \setminus \{u_i, u_{i+1}, \cdots, u_{j-1}\}$. Let v' be another crossbar in stage i. If v' differs from v in a bit in I, then clearly, $V_j(v') \cap V_j(v) = \phi$; if not, then $V_j(v') = V_j(v)$. Since i, j, v, v' are arbitrary, the network is in B. $\square$

Theorem 1.5.13. $BP \subset d^P B$.

Proof. That $BP \subseteq d^P B$ follows from Corollary 1.5.11 and Theorem 1.5.12. That the containment is strict follows from Fig. 1.5.6. $\square$

Theorem 1.5.14. *A $d^P B$ network is equivalent to a BP network.*

Proof. We prove Theorem 1.5.14 by induction on the number of stage. Consider an s-stage $d^P B$ network.

(i) $s = 2$. Suppose v of stage 1 is connected to the set $V_2(v)$. Let v be another crossbar in stage 1 and connected to $w \in V_2(v)$. By the buddy property, $V_2(w) = V_2(v)$. Since there are $d - 1$ choices of v', these v' together with v form a $d \times d$ complete bipartite graph $K_{d,d}$ with $V_2(v)$. Furthermore, $V_2(v') \cap V_2(v) = \phi$ for any $v'' \in v \cup \{v'\}$. Since v is arbitrary, G_{12} consists of $d^{n-1} K_{d,d}$ whose equivalence to a BP network is clear.

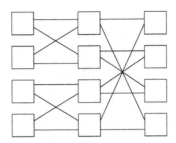

Figure 1.5.6: A $d^P B$ network which is not a BP-network

(ii) $s = 3$. By the d^P property, the network has d^{n-k} components for some $1 \leq k \leq n$. Recall that from (1) the subgraphs G_{12} and G_{13} must each consist of $d^{n-1} K_{d,d}$. For $k = 1$, then no two $K_{d,d}$ in G_1 can be connected through G_2. Therefore G_{13} must consist of d^{n-1} copies of concatenation of two $K_{d,d}$, with the outputs of the former identified with the inputs of the latter (see Fig. 1.5.7).

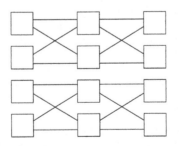

Figure 1.5.7: Concatenation of $K_{2,2}$

Clearly, G_{13} is equivalent to a BP network.

For $k = 2$, first suppose G_{13} is obtained by connecting each d-set $D = \{D_1, \cdots, D_d\}$, when each D_i is a $K_{d,d}$ in G_1, into one component in G_{13}. Note that the connection is done by a d-set $D' = \{D'_1, \cdots, D'_d\}$ of $K_{d,d}$ in G_2. If two crossbars of the same D_i are connected to a D'_j, then one member of $D \setminus D_i$ will not be connected to D'_j, violating the buddy property. Therefore, the d crossbars in a D_i must go to distinct D'_j, or all D'_j. Since we can permute the stage-2 crossbars in a D arbitrarily, and independently for each D, the stage-2 crossbars in each D can be ordered such that the kth one goes to the $k^{th} D'$, which is clearly a BP network. Figure 1.5.8 illustrates how to permute.

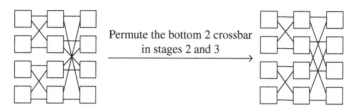

Figure 1.5.8: A permutation to achieve BP

Suppose G_{13} is obtained otherwise. There must exist a d'-set of $K_{d,d}$, $d' > d$, in G_1 connected in G_2 through a d'-set of $K_{d,d}$ in G_2. Note that an input in this component touches only d^2 among the dd' outputs. Hence there must exist another input reaching some, but not all, of these d^2 outputs, violating the buddy property.

For $k \geq 3$, then the situation described in the last paragraph must also happen.

(iii) $s \geq 4$. Consider the two subnetworks G_{13} and G_{2s}. By induction, G_{13} can be represented by a vector (u_1, u_2) and G_{2s} by $(u'_1, \cdots, u'_{s-2})$. By Corollary 1.5.6, we can permute the crossbars in stage k, $2 \leq k \leq s$, such that $u'_1 = u_2$ and $u'_k = u''_k$ for $2 \leq k \leq s-2$. Therefore G_{1s} is represented by the vector $(u_1, u_2, u''_3, \cdots, u''_{s-1})$, i.e., G_{1s} is a BP network.

$\square$

Corollary 1.5.15. *Two $d^P B$ networks are equivalent if the characterization vector of one can be obtained from the other through a permutation.*

Note that Corollary 1.5.15 generalizes Theorem 1.5.1.

Figure 1.5.9(a) gives a d^P network which is not equivalent to a BP network (since it has a component of size $\geq d^2$). Therefore the buddy condition can not be dropped from Theorem 1.5.13.

Figure 1.5.9(a) is also an example of a d^P network which is not equivalent to a buddy network, while Fig. 1.5.9(b) gives a buddy network which is not equivalent to either a d^P or a BP network.

Finally, we have

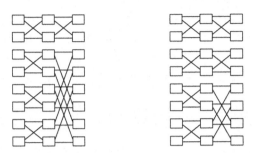

Figure 1.5.9: A 2^P network and a buddy network

Theorem 1.5.16. *The number of equivalent classes among s-stage bit-permutation networks is* $\sum_{t=1}^{n-1} \sum_{i=1}^{t} \left(\frac{1}{t!}\right) (-1)^{t-i} \binom{t}{i} i^{s-1}$.

Proof. The number of canonical sequences of length m with t distinct elements is simply the number of ways of labeling m cells with exactly t symbols. Using the principle of inclusion and exclusion, this number is

$$\sum_{i=1}^{t} (-1)^{t-i} \binom{t}{i} i^m \ .$$

Let $i_1, \ldots, i_t$ denote the ordering of the t symbols according to their appearances in the m cells. Then the $t!$ possible orderings reduce to the same canonical sequence, which has the ordering $(1, \ldots, t)$. Therefore we have to divide by $t!$ to obtain the number of canonical sequences with exactly t symbols. Summing over t and setting $m = s - 1$, we obtain Theorem 1.5.16. $\qquad\square$

Note that the number of equivalent classes is independent of d.

For $B(0, n)$, Li also introduced a notion called guide to facilitate the self-routing. A *guide* is an $n \times n$ upper-left triangular matrix(g_i) where the set $\{1, ..., n\}$ labels the rows (from top to bottom) and the columns (from left to right). Then $g_{ii} = n$ for $1 \leq i \leq n$, and g_{ij} represents the number that $g_{i,j-1}$ permutes to Π_{j+1}. For example, for SE_4, $\Pi_i = (1432)$ for all i. Then $g_{12} = 3$ since 4 permutes to 3 in Π_3. $g_{13} = 2$ since 3 permutes to 2 in Π_2, and so on. The guide is shown in Fig. 1.5.10(a). For BY_4^{-1}, $\Pi_i = (i, 4)$ for $1 \leq i \leq 3$; its guide is shown in Fig. 1.5.10(b). For BL_4, $\Pi_i = (i, i+1, ..., n)$; its guide is shown in Fig. 1.5.10(c). For BY_4, $\Pi_i = (n-i, 4)$ for $1 \leq i \leq 3$; its guide is shown in Fig. 1.5.10(d).

$$
\begin{array}{cccc}
4\ 3\ 2\ 1 & 4\ 1\ 1\ 1 & 4\ 1\ 1\ 1 & 4\ 3\ 3\ 3 \\
4\ 3\ 2 & 4\ 2\ 2 & 4\ 2\ 2 & 4\ 2\ 2 \\
4\ 3 & 4\ 3 & 4\ 3 & 4\ 1 \\
4 & 4 & 4 & 4 \\
(a) & (b) & (c) & (d)
\end{array}
$$

Figure 1.5.10: Guides for SE_4, BY_4, BL_4 and BY_4^{-1}

To route $(x_1, ..., x_n)$ to $(y_1, ..., y_n)$ in a BP network represented by the canonical vector $(u_1, ..., u_{n-1})$, set $x_n = y_{g_{in}}$ at the stage-i switch, i.e., the path goes to the upper (lower) output if $y_{g_{in}} = 0(1)$. Π_i, $1 \leq i \leq n-1$, then moves $y_{g_{in}}$ from position g_{ij} to position $g_{i,j+1}$. So at stage n, $y_{g_{in}}$ is moved to position g_{in}. Namely, the outcome vector is $(y_1, ..., y_n)$. We illustrate by some examples.

$$
SE_4 : (x_1x_2x_3x_4) \overrightarrow{g_{14}} (x_1x_2x_3y_1) \overrightarrow{\Pi_1} (x_2x_3y_1x_1) \overrightarrow{g_{24}} (x_2x_3y_1y_2) \overrightarrow{\Pi_2}
$$
$$
(x_3y_1y_2x_2) \overrightarrow{g_{34}} (x_3y_1y_2y_3) \overrightarrow{\Pi_3} (y_1y_2y_3x_3) \overrightarrow{g_{44}} (y_1y_2y_3y_4).
$$

$$
BY_4^{-1} : (x_1x_2x_3x_4) \overrightarrow{g_{14}} (x_1x_2x_3y_3) \overrightarrow{\Pi_1} (x_1x_2y_3x_3) \overrightarrow{g_{24}} (x_1x_2y_3y_2) \overrightarrow{\Pi_2}
$$
$$
(x_1y_2y_3x_2) \overrightarrow{g_{34}} (x_1y_2y_3y_1) \overrightarrow{\Pi_3} (y_1y_2y_3x_1) \overrightarrow{g_{44}} (y_1y_2y_3y_4).
$$

1.6 Graphs and Channel Graphs

A graph G consists of a set V of *vertices* and a set E of pairs of vertices called *edges*. If the pairs are ordered pairs, then a graph is known as a *digraph*, a vertex is called a *node* and an edge an *arc*. The degree of a vertex v, $d_G(v)$, is the number of edges it is incident to. Two vertices (edges) are *adjacent* if they are incident to the same edge (node). A *coloring* (an *edge-coloring*) of a graph is to assign a color to each node (edge) such that all adjacent nodes (edges) have different colors. We say that the graph can be c-(edge-)colored if c colors suffice. These terms have their counterparts for digraphs. In particular, the modifier "in" or "out" is used to associate with arcs coming into or going out from a node.

In a weighted graph, each edge has a weight assumed to be a fraction. An edge-coloring of a weighted graph satisfies the requirement that the sum of weights of all edges of the same color incident to a vertex v must not exceed 1 for all v.

By treating a crossbar as a node and a link as an arc, a switching network is very much like a digraph except that each input (output) switch has external links dangling without connecting to any nodes and hence cannot be considered as arcs (but the computer network version does not have this problem). To remedy this irregularity, the graph theorist prefers to define a true digraph $G(V, I, O, E)$ from a network by converting each link as a node including the inputs I and the outputs O, while a crosspoint connecting two links in the network becomes an arc in the graph. Note that a crossbar is represented by a complete bipartite subgraph whose recognizability may depend on the drawing of $G(V, I, O, E)$. Figure 1.6.1 shows the digraph representation of the network in Fig. 1.2.1.

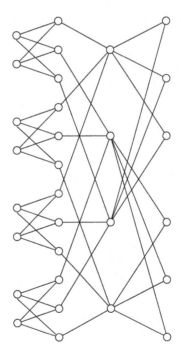

Figure 1.6.1: Digraph of Fig. 1.2.1.

While a telephone network does have the external links to connect to outside world, a processor-memory network can use the input crossbars to represent processors, the output crossbars to represent memories, so that requests are internally generated. Such a network is often drawn without the external links, and itself interpreted as a digraph.

A graph is called *bipartite* if its vertices can be partitioned into two disjoint parts such that edges exist only between parts. A frame F can be modelled by a bipartite graph $G(F)$ with the input crossbars and output crossbars as the two parts, and an edge (X, Y) for each request (x, y) with x an input on X and y an output on Y. For a 3-stage Clos network, the requirement that only one request from each input crossbar and output crossbar can be connected through a given middle crossbar can then be translated into an edge-coloring of $G(F)$ where each color corresponds to a middle crossbar. Namely, we can assign each middle crossbar with a distinct color which will route all edges of that color.

Bipartite graphs are also used to describe the connection pattern between two adjacent stages. We define some bipartite graphs with special properties useful to nonblocking networks. Let N denote the number of inputs and M the number of outputs.

Expander. For a given family F of input-subsets, each input-subset S can reach an output subset S' with $|S'| > |S|$. Depending on different constraints on F and different requirements on $|S'|$, various expanders can be defined. The original definition is: An (N, d, p)-expander is a d-regular graph with $M = N$ such that every set of x inputs, $1 \le x \le N$, has at least $x + p(1 - x/N)x$ outputs as neighbors. Pinsker (1973) first proved the existence of a linear-cost (N, d, p)-expander and Margulis (1975) gave the first construction but leaving p unspecified. Gabber–Galil (1981) provided this constant. The currently best construction requires a cost about $30N$.

In a multiplicative (α, β)-expander, each subset S, $|S| \le \alpha N$, of inputs can reach $\beta|S|$ outputs with $\beta > 1$. In an additive (p, q)-expander, each subset S, $|S| \le p$, of inputs can reach $p + q$ outputs, $q > 0$.

Concentrator. For any k-subset K_1 of inputs, there exists a matching between K_1 and a k-subset K_2 of outputs (K_2 is not prespecified) as long as $k \le \min\{N, M\}$. Pinsker (1973) showed that a concentrator can be constructed from an expander with essentially the same cost.

Partial concentrator. A concentrator becomes a partial concentrator, denoted by (N, M, c), when the above condition holds only for $k \le$ a constant c. c is called the *capacity* of the partial concentrator.

Superconcentrator. A concentrator with K_2 also specified (but note that the matching is not specified). Valiant (1976) first proved the existence of a linear-cost superconcentrator. Gabber–Galil gave a method to convert an expander to a superconcentrator with the same cost complexity. The currently best construction requires a cost about $100N$.

Hyperconcentrator. A superconcentrator with a consecutive set K_2.

Infra concentrator. The set of the first k inputs can be mapped to some set of k outputs preserving the order.

In particular

Infra connector. The set of the first k inputs can be mapped to any set of k outputs prescribing the order.

The reader is referred to Alon–Galil–Milman (1987), Tanner (1987), Pippenger (1990) for more detailed discussions on these bipartite graphs.

The shuffled $BY^{-1}(n)$ differs from $BY^{-1}(n)$ in the fact that inputs are assigned to the input crossbars in a shuffled pattern, i.e., suppose the N inputs are lined up at "stage 0", then inputs i and $i + N/2$ are connected to input crossbar i for $1 \leq i \leq N/2$. Gfman (1978) proved

Theorem 1.6.1. *The shuffled $BY^{-1}(n)$ is an infra connector.*

Proof. The restrictions on routable permutations for a shuffled $BY^{-1}(n)$ is that if inputs i and j have the same last l bits, then $\Pi(i)$ and $\Pi(j)$ cannot have the same first $n - l$ bits. On the other hand, when inputs are consecutive and the mapping to output is monotone, then $i < j$ implies $\Pi(j) - \Pi(i) \geq j - i$. Hence if i and j have the same last l bits, then $\Pi(i) - \Pi(j) \geq 2^l$, i.e., $\Pi(i)$ and $\Pi(j)$ cannot have the same first $n < l$ bits. Consequently, such a permutation is routable by the shuffled $BY^{-1}(n)$. $\square$

For a given MIN ν, the channel graph $CG(x, y)$ between an input x and an output y is the union of all paths connecting x and y. The channel graph, first proposed by Lee (1955) and LeGall (1956) independently, has been widely used in studying the blocking probability of MIN. Recently, Shyy and Lea (1991) showed that it can also be useful in the study of nonblocking networks.

Consider an s-stage channel graph CG. For odd s, CG is called a *double-tree channel graph* if CG can be partitioned at some stage into two trees whose leaves are identified at that stage. For even s, there exists a stage of links connecting the identified pairs of leaves. A double-tree is *series parallel* if the leaves of the two trees in their natural orders are mapped by an identity mapping; otherwise, it is *spiderweb*. Figure 1.6.2 illustrates both types. A symmetric *s*-stage series-parallel channel graph can be sequentially decomposed into $\lfloor (s-1)/2 \rfloor$ *shells* where shell i consists of links between stages i and $i+1$, and between stages $s-i$ and $s-i+1$.

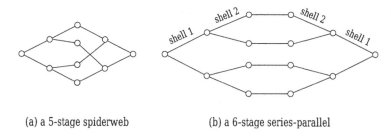

(a) a 5-stage spiderweb   (b) a 6-stage series-parallel

Figure 1.6.2: Two double-trees

Clearly, $CG(x, y)$ plays an important role in determining whether the (x, y) request can be connected. It is particularly convenient if $CG(x, y)$ is invariant to x and y, since then the connectability of one pair stands for the whole network. Often, the hardware of a network, namely, the number of stages, the number and the size of the crossbars at each stage, is given. The question is to determine the interconnection pattern inducing the best channel graph. It is well known (Hwang, 1979) that a spiderweb channel graph is usually better than a series-parallel channel graph for blocking networks. Among series-parallel channel graphs with fixed number of stages and fixed number of paths, it is better to have branching at an outer shell than an inner shell. It is of interest to note that in Shyy and Lea's analysis of channel graph for nonblockingness (see Sec. 2.2), it is also better to have more branching at outer stages, even though the two underlying models are totally different.

Since the nonblockingness analyses of bit permutation networks to be discussed in later chapters depend only on the (invariant) channel graphs of the networks, the question arises whether nonequivalent bit permutation networks can have isomorphic channel graphs. If that were the case, then the study of equivalent bit permutation networks becomes less relevant. Tong, Hwang and Chang (1998) gave the reassuring "no" answer. First they proved

Theorem 1.6.2. *The channel graph of a connected bit permutation network is invariant to x and y.*

Proof. Let $CG(x)$ denote the union of all paths from an input x. Let $V_i(x)$ denote the set of stage-i nodes in $CG(x)$. Then $V_i(x)$ consists of those nodes which may differ from (the d-nary representation of) x in bits $I_j = j = 1, \ldots, i$. Therefore $CG(x)$ is invariant to x. Let $x' \neq x$ be another input. Since the bit permutation network is connected, $G(x)$ intersects $G(x')$ at some stage k. By the buddy property $V_k(x) = V_k(x')$. Therefore $CG(x, y)$, being the concatenation of the graph from x to $V_k(x)$ and the graph from $V_k(x)$ to y is isomorphic to the concatenation of the graph from x' to $V_k(x')$ and the graph from $V_k(x')$ to y, which is $CG(x', y)$. Similarly, we can prove $CG(x, y) = CG(x, y')$. Hence $CG(x, y) = CG(x', y) = CG(x', y')$. $\qquad\square$

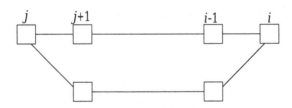

Figure 1.6.3: A cycle from stage j to stage i

Theorem 1.6.3. *Two connected bit permutation networks are isomorphic if and only if their channel graphs are isomorphic.*

Proof. The "only if" part is trivial. We prove the "if" part.

Let $\nu_1(n; I_{k_1}, \ldots, I_{k_{s-1}})$ and $\nu_2(n; I_{k'_1}, \ldots, I_{k'_{s-1}})$ denote two nonisomorphic s-stage connected bit permutation networks. Then there exists a smallest stage i such that $k_i \neq k'_i$. Without loss of generality, assume $k_i > k'_i$. Then $k'_i \in \{k'_1, \ldots, k'_{i-1}\} = \{k_1, \ldots, k_{i-1}\}$. Let j be the largest subscript such that $k'_i = k'_j = k_j$. Then the channel graph of ν_2 from stage j to i contains a cycle as shown in Fig. 1.6.3, which is not present in the channel graph of ν_1. $\qquad\square$

References

Ackroyd, M. H. 1979. Call repacking in connecting networks. *IEEE Trans. Commun.*, **27**, 589–591.

Agrawal, D. P. 1983. Graph theoretical analysis and design of multistage interconnection networks. *IEEE Trans. Comput.*, **C32**, 637–648.

Alon, N., Galil, Z., & Milman, V. D. 1987. Better expanders and superconcentrators. *J. Alg.*, **8**, 337–347.

Beneš, V. E. 1965. *Mathematical Theory of Connecting Networks and Telephone Traffic*. New York: Academic.

Bermond, J. C., Fourneau, J. M., & Jean-Marie, A. 1987. Equivalence of multistage interconnection networks. *Inform. Proc. Lett.*, **26**, 45–50.

Cantor, D. G. 1971. On nonblocking switching networks. *Networks*, **1**, 367–377.

Chang, G. J., Hwang, F. K., & Tong, L. D. 1999. Characterizing bit permutation networks. *Networks, 33, 261-267.*

Clos, C. 1953. A study of non-blocking switching networks. *Bell Syst. Tech. J.*, **32**, 406–424.

Gabber, O., & Galil, Z. 1981. Explicit construction of linear size superconcentrators. *J. Comput. Syst. Sci.*, **22**, 407–420.

Halpenny, L. 1990. Nonblocking staged and repeated stage networks. Unpublished manuscript.

Hui, J. Y. 1990. *Switching and Traffic Theory for Integrated Broadband Networks*. Norwell, MA: Kluwer.

Hwang, F. K. 1979. Superior channel graphs. *In: Proc. 9th Int. Teletraffic Cong.* Torremolinos, Spain.

Hwang, F. K. 2003. Equivalence among extra-stage banyan-type networks, preprint.

Hwang, F. K., Liaw, S. C., & Yeh, H. G. 1998. Equivalent classes of extra-stage networks, unpublished.

Kruskal, C. P., & Snir, M. 1986. A unified theory of interconnection network structure. *Theo. Comput. Sci.*, **48**, 75–94.

Lawrie, D. H. 1976. Access and alignment of data in an array processor. *IEEE Trans. Comput.*, **25**, 1145–1155.

Lee, C. Y. 1955. Analysis of switching networks. *Bell Sytem Tech. J.*, **34**, 1287–1315.

LeGall, P. 1956. Étude du blocage dans les systèmes de commutation téléphonique. *Ann. Télécommun.*, **11**.

Li, S.-Y. R. 2001. Algebraic Switching Theory and Broadband Applications. *New York: Academic.*

Margulis, G. A. 1975. Explicit constructions of concentrators. *Prob. Inform. Transm.*, **9**, 325–332.

Ofman, Y. P. 1965. A universal automation. *Trans. Moscow Math. Soc.*, **14**, 200–215.

Parker, D. S. 1980. Notes on shuffle/exchange-type networks. *IEEE Trans. Comput.*, **29**, 213–222.

Perrier, P., & Prucnal, P. 1989. Self-clocked optical control of a self-routed photonic switch. *J. Lightwave Tech.*, **7**.

Pinsker, M. 1973. On the complexity of a concentrator. *Proc. 7th ITC,* Stockholm, Sweden, 318/1–318/4.

Pippenger, N. 1990. Communication networks. *Pages 807–833 of:* Leeuwen, J. V. (ed), *Handbook of Theoretical Computer Science.* Amsterdam: Elsevier.

Shyy, D.-J., & Lea, C.-T. 1991. $Log_2(N, m, p)$ strictly nonblocking networks. *IEEE Trans. Commun.*, **39**, 1502–1510.

Siegel, H. J., & Smith, S. D. 1978. Study of multistage SIMD interconnecting networks. *Pages 307–314 of: Proc. 5th Ann. Symp. Comput. Arch.*

Smyth, C. Y. 1988. Nonblocking photonic switch networks. *IEEE J. Select. Areas Commun.*, **6**, 1052–1062.

Spanke, R. A., & Beneš, V. E. 1987. N-stage planar optical permutation network. *Appl. Opt.*, **26**, 1226–1229.

Tanner, R. M. 1987. Explicit concentrators from generalized N-gons. *SIAM J. Alg. Disc. Method,* **5**, 287–293.

Tong, L. D., Hwang, F. K., & Chang, G. J. 1998. Channel graphs of bit permutation networks, Theor. Comput. Sci., **263**, 139-143.

Thompson, C. D. 1978. Generalized connection network for parallel processor interconnection. *IEEE Trans. Comput.*, **27**, 1119–1124.

Valiant, L. G. 1976. Graph-theoretic properties in computational complexity. *J. Comput. Syst. Sci.*, **13**, 278–285.

Varma, A., & Raghavendra, C. S. (eds). 1994. *Interconnection Networks for Multiprocessors and Multicomputers: Theory and Practice.* IEEE Computer Soc., Los Alamitos, CA.

Wu, C.-L., & Feng, T.-Y. 1980. On a class of multistage interconnection networks. *IEEE Trans. Comput.*, **29**, 694–702.

Chapter 2. Nonblocking Networks

2.1 3-stage Clos Network

The network $[X_{n_1m}, X_{r_1r_2}, X_{mn_2}]$ is often called a 3-stage Clos network to celebrate the following theorem by Clos (1953) who started the study of nonblocking networks. A 3-stage Clos network is also denoted by $C(n_1, r_1, m, n_2, r_2)$, and in the symmetric case, by $C(n, m, r)$. Let I_i denote the ith input crossbar, O_i the ith output crossbar and M_i the ith crossbar in the middle stage.

Theorem 2.1.1. *Assuming* $\min\{r_1, r_2\} \geq 2$, $[X_{n_1m}, X_{r_1r_2}, X_{mn_2}]$ *is SNB if and only if* $m \geq m^o$, *where*

$$m^o = \min\{n_1 + n_2 - 1, \ N_1, \ N_2\}.$$

Proof. Without loss of generality, assume the new request is from I_1 to O_1. Clearly, $m = N_1$ or N_2 is sufficient since only $\min\{N_1, N_2\}$ requests can be generated, and we can route them each through a distinct M. Furthermore, if the busy co-inlets and co-outlets are each routed through a distinct M, then at most $(n_1 - 1) + (n_2 - 1)$ M's are taken; so $n_1 + n_2 - 1$ M's also suffice.

Next we prove necessity. Suppose $m < m^o$. Then the co-inlets can block as many as $\min\{n_1 - 1, N_2 - n_2\}$ M's, and the co-outlets can block as many as $\min\{n_2 - 1, N_1 - n_1\}$ M's, disjoint from the first set. Since $\min\{n_1 - 1, N_2 - n_2\} + \min\{n_2 - 1, N_1 - n_1\} = m^o - 1 \geq m$, the new request is blocked. $\qquad\square$

Corollary 2.1.2. *For the above network, the cost is*

$$m^0 \cdot [N_1 + (N_1/n_1)(N_2/n_2) + N_2].$$

In the symmetric case, set $n = (N/2)^{1/2}$. Then the cost $\leq 4\sqrt{2}N^{3/2}$.

Much less is known for WSNB; perhaps because it is not easy to come up with intelligent routing algorithms which can make a difference. Some possible candidates are:

save-the-unused (STU). Do not route through an empty M_j unless there is no choice.

packing (P). Route through anyone of the busiest M's.

minimum index (MI). Route through the M_j with the smallest index if possible.

cyclic dynamic (CD). If M_j is used in routing the last request, try M_{j+1}, $M_{j+2}, \ldots$ in that cyclic order.

cyclic static (CS). Same as CD except starting from M_j.

Note that STU includes P. Also note that STU and P in fact represent two classes of algorithms since there are various ways of breaking ties. A network is WSNB under A if it is nonblocking under any tie-breaking choice. The existence of a WSNB network was first demonstrated by Beneš (1965) on $C(n, m, r)$.

Theorem 2.1.3. $C(n, m, 2)$ is WSNB under STU if $m \geq \lfloor 3n/2 \rfloor$.

Proof. Let s be a state and let $y(s)$ denote the number of middle crossbars carrying at least one connection. Let $y_i(s)$ denote the number of middle crossbars carrying a set T_i of connections where $T_1 = \{(I_1, O_1)\}$, $T_2 = \{(I_1, O_2)\}$, $T_3 = \{(I_2, O_1)\}$, $T_4 = \{(I_2, O_2)\}$, $T_5 = \{(I_1, O_1), (I_2, O_2)\}$ and $T_6 = \{(I_1, O_2), (I_2, O_1)\}$. We will prove the theorem by induction on the number of steps to reach s (from the initial empty state):

(i) $y(s) \leq \lfloor 3n/2 \rfloor$,

(ii) $y_1(s) + y_4(s) + y_5(s) \leq n$,

(iii) $y_2(s) + y_3(s) + y_6(s) \leq n$.

All three claims are trivially true at step 1. Consider a general step from state s' to s. If s is obtained from s' by deleting a connection, the three claims obviously remain true. So assume s is obtained from s' by adding a connection. Without loss of generality, assume it is the (I_1, O_1)-connection. If $y_4(s') > 0$, then a crossbar carrying an (I_2, O_2)-connection can carry the new request. Thus $y_4(s) = y_4(s') - 1$, $y_5(s) = y_5(s') + 1$ and the claims remain true. Therefore, we assume $y_4(s') = 0$.

(i) Since I_1 and O_1 can each be engaged in at most $n - 1$ connections,

$$y_1(s') + y_2(s') + y_5(s') + y_6(s') \leq n - 1,$$
$$y_1(s') + y_3(s') + y_5(s') + y_6(s') \leq n - 1.$$

Using the induction hypothesis (iii)

$$y_2(s') + y_3(s') + y_6(s') \leq n.$$

Adding up,

$$2\left[y_1(s') + y_2(s') + y_3(s') + y_5(s') + y_6(s')\right] \leq 3n - 2,$$

or

$$y(s') \leq \lfloor 3n/2 \rfloor - 1.$$

Route the new request through an unused middle crossbar. Then $y(s) \leq \lfloor 3n/2 \rfloor$.

(ii) $\begin{aligned} y_1(s) + y_4(s) + y_5(s) &= 1 + y_1(s') + y_4(s') + y_5(s') \\ &= 1 + y_1(s') + y_5(s') \leq n, \end{aligned}$

since every connection involves I_1.

(iii) $y_2(s) + y_3(s) + y_6(s) = y_2(s') + y_3(s') + y_6(s') \leq n.$

If the new call is (I_1, O_2) or (I_2, O_1), then Eq. (ii) instead of (iii) will be critically used. □

The first proof of Theorem 2.1.3 was credited to E. F. Moore by Kurshan–Beneš (1980). Note that the assumption that (I_1, O_1) engages an unused switch only when $y_4(s') = 0$ observes the STU routing.

Smith (1977) proved the condition in Theorem 2.1.3 is also necessary by proving

Theorem 2.1.4. $C(n, m, r)$ *is not WSNB under either P or MI if* $m < 2n - \lfloor n/r \rfloor$.

Instead of giving a proof, we will state and prove a stronger result.

For large r it is more convenient to represent a state by an $r \times r$ matrix where the rows are labeled by input crossbars, the columns by output crossbars, and cell (i, j) consists of the set M^{ij} of middle crossbars each carrying a call from I_i to O_j. The stronger result holds for those algorithms for which $M^{11} = M^{22} = \cdots = M^{rr}$, $|M^{ii}| = n$, is a state, called the *n-uniform state*. All the algorithms we mentioned above can reach the n-uniform state from the empty state. For STU, P, MI and CS, let the request sequence be n iterations of the subsequence $(I_1, O_1), (I_2, O_2), \ldots, (I_r, O_r)$. Then all requests in the jth iteration are routed through M_j. For CD, we can force a request to be carried by a desirable middle crossbar by repeated generating and canceling the request until it is assigned to the right crossbar (each iteration moves the assignment to the next crossbar). Du–Fishburn–Gao–Hwang (2001) proved

Theorem 2.1.5. *Let A be an algorithm which has the n-uniform state. Then $C(n, m, r)$ is WSNB under A only if $m \geq 2n - \lceil n/2^{r-1} \rceil$.*

Proof. Consider the n-uniform state. Without loss of generality, assume $M^{ii} = \{M_1, \ldots, M_n\}$. In the operation to be described, M^{11} changes after every move. Define S_j to be the intersection of M^{jj} with the current M^{11}. At the beginning $|S_2| = n$. Let S_2' and S_2'' be two disjoint $\lfloor |S_2|/2 \rfloor (= \lfloor n/2 \rfloor)$-subsets of S_2. Delete S_2' from M^{11}, delete S_2'' from M^{22}, and add $\lfloor n/2 \rfloor$ new (I_1, O_2) connections, which must be carried by M_j's with $j > n$, say, by $M_{n+1}, \ldots, M_{n+\lfloor n/2 \rfloor}$. Theorem 2.1.5 is verified for $r = 2$.

Update M^{11}. Then $|S_3| = \lceil n/2 \rceil$. Again, let S_3' and S_3'' be two disjoint $\lfloor |S_3|/2 \rfloor$-subsets of S_3. Delete S_3' from M^{11}, delete S_3'' from M^{33}, and add $\lfloor |S_3|/2 \rfloor$ new (I_1, O_3) connections which must be carried by M_j's with $j > n + \lfloor n/2 \rfloor$, say, by $M_{n+\lfloor n/2 \rfloor+1}, \ldots, M_{n+\lfloor n/2 \rfloor+\lfloor |S_3|/2 \rfloor}$. It is easily verified that $\lfloor n/2 \rfloor + \lfloor \lceil n/2 \rceil /2 \rfloor = n - \lfloor n/4 \rfloor$. Theorem 2.1.5 is verified for $r = 3$.

In general, at the $(j-1)^{\text{st}}$ move, we delete S_j' from M^{11}, delete S_j'' from M^{jj}, and add $\lfloor |S_j|/2 \rfloor = \lfloor \lceil n/2^{j-2} \rceil /2 \rfloor$ new Ms. Again, it is easily verified that $2n - \lceil n/2^{r-2} \rceil + \lfloor \lceil n/2^{r-2} \rceil /2 \rfloor = 2n - \lceil n/2^{r-1} \rceil$. □

The following example illustrates the case $n = 4$ and $r = 3$. Define $[i, j] = \{M_i, M_{i+1}, \cdots, M_j\}$ for $i \leq j$.

[1,4]		
	[1,4]	
		[1,4]

$\rightarrow$

[1,2]		
	[3,4]	
		[1,4]

$\rightarrow$

[1,2]	[5,6]	
	[3,4]	
		[1,4]

$\rightarrow$

[1]	[5,6]	
	[3,4]	
		[2,4]

$\rightarrow$

[1]	[5,6]	[7]
	[3,4]	
		[2]

For P and $r \geq 3$, Yang-Wang (1999) proved the stronger necessary condition $m \geq 2n - \lceil n/F_{2n-1} \rceil$, where F_k is the kth Fibonacci number by analytically solving a linear programming problem. Du-Fishburn-Gao-Hwang (2001) proved the necessary and sufficient condition $m \geq 2n-1$ by a quite complicated argument. Here we adopt a more elementary proof of Chang-Guo-Hwang-Lin (2004).

Theorem 2.1.6. *For P and STU, $C(n,m,r)$, $r \geq 3$, is WSNB if and only if $m \geq 2n - 1$.*

Proof. The "if" part is trivial since $2n - 1$ guarantees SNB. We prove the "only if" part by showing that for $r = 3$ there exists a sequence of calls and disconnections forcing the use of $2n - 1$ middle switches under either P or STU:

$$
\begin{array}{c|c}
[1,n] & \\
\hline
& [1,n]
\end{array}
\rightarrow
\begin{array}{c|c}
& n \\
\hline
n+1 & [1,n-1]
\end{array}
\rightarrow
\begin{array}{c|c|c}
& n & n+1 \\
\hline
n+1 & [1,n-1] &
\end{array}
$$

$$
\rightarrow
\begin{array}{c|c}
n & n+1 \\
\hline
& [1,n-1] \\
n+1 &
\end{array}
\rightarrow
\begin{array}{c|c|c}
n & & n+1 \\
\hline
n+2 & [1,n-1] & \\
n+1 & &
\end{array}
$$

$$
\rightarrow
\begin{array}{c|c}
n & [n+1,n+2] \\
\hline
n+2 & [1,n-1] \\
n+1 &
\end{array}
$$

$$
\rightarrow
\begin{array}{c|c}
n & [n+1,n+2] \\
\hline
& [1,n-1] \\
[n+1,n+2] &
\end{array}
$$

$$
\rightarrow \cdots \rightarrow
\begin{array}{c|c}
n & [n+1,2n-2] \\
\hline
& [1,n-1] \\
[n+1,2n-2] &
\end{array}
$$

$$
\rightarrow
\begin{array}{c|c}
n & [n+1,2n-2] \\
\hline
2n-1 & [1,n-1] \\
[n+1,2n-2] &
\end{array}
$$

$\square$

Since adding input and output switches does not decrease m, the proof for $r = 3$ works for all larger r.

Chang-Guo-Hwang-Lin also obtained the necessary and sufficient conditions for CD, CS and MI. First, two lemmas.

Lemma 2.1.7. *CD can reach any state s from any state s'.*

Proof. Since we can disconnect all paths in s' to reach the empty state, it suffices to prove for s' the empty state. We prove this by adding each M_k in s to its proper cell one by one. Suppose M_k is in cell (i,j). Consider a pair

(I, O). Suppose CD assigns M_h to connect the pair. If $h \neq k$, disconnect the pair and regenerate it immediately. Then CD would assign M_{h+1} to connect the pair. Repeat this until M_k is assigned. Since M_k is arbitrary, s can be reached.   □

For CS we prove a weaker property.

Lemma 2.1.8. *Let state s be obtained from s' by adding $[i, j]$, $i < j$, to a cell C. Then s can be reached from s' under CS.*

Proof. Suppose the last assignment is M_k in s'. Since $i < j$, we can add at least two connections in C. Then M_k and M_{k+1} will be assigned. If $k \neq i$, disconnect the connection through M_k and regenerate a connection in C, for which M_{k+2} will be assigned. Continue this until M_i and M_{i+1} are assigned. Then add $j - i - 1$ connections to C for which $M_{i+2}, \ldots, M_j$ will be assigned.   □

Theorem 2.1.9. $C(n, m, r)$ *for $r \geq 2$ is WSNB under CD and CS if and only if $m \geq 2n - 1$.*

Proof. The "if" part is trivial since $C(n, 2n - 1, r)$ is SNB, hence WSNB. To prove the "only if" part, we claim that if $m = 2n - 2$, then there exists a blocking state.

It suffices to prove for the minimum r which is 2 here. By Lemmas 2.1.7 and 2.1.8, the state in which cell $(1, 1)$ contains $[1, n - 1]$ and cell $(2, 2)$ contains $[n, 2n - 2]$ can be reached. But a new pair $(1, 2)$ is blocked.   □

For the MI case, first, a lemma :

Lemma 2.1.10. *Consider a state s in $C(n, m, 2)$ consisting of x (I_1, O_1) calls carried by the set X of middle switches, and y (I_2, O_2) calls carried by the set Y of middle switches such that $X \cap Y = \phi$, $X \cup Y = \{1, \cdots, x + y\}$. Then a state s' can be obtained from s, where s' is same as s except that x becomes x', and y becomes $y' = x + y - x'$.*

Proof. Without loss of generality, assume $x' > x$ (otherwise we work with y). Delete $x' - x$ (I_2, O_2) calls carried by $S = \{$*the smallest $x' - x$ indices in Y*$\}$ from s. Add $x' - x$ new (I_1, O_1) requests. By the MI rule, these new requests must be carried by S. Thus s' is obtained.   □

Theorem 2.1.11. $C(n, m, r)$ *for $r \geq 2$ is WSNB under MI if and only if $m \geq 2n - 1$.*

Proof. Again, it suffices to prove that $m = 2n - 1$ is necessary for WSNB for $r = 2$. Theorem 2.1.11 is easily verified for $n = 2$. We prove the general $n \geq 3$ case by induction on n.

By induction, $m = 2n - 3$ is necessary for $C(n - 1, m, 2)$ to be WSNB. Therefore there exists a state s

$$\begin{array}{c|c} X & 2n - 3 \\ \hline & Y \end{array}$$

in $C(n, 2n - 1, 2)$.

Such that $x = y = n - 2$, $X \cap Y = \phi$, $X \cup Y = \{1, \cdots, 2n - 4\}$. Therefore we can obtain a state s' from s by adding $2n - 2$ to the $(1, 2)$ cell. Delete the four calls carried by $[2n - 7, 2n - 4]$ in the $(1, 1)$ and $(2, 2)$ cells, and use Lemma 2.1.10 to rebalance the members of calls carried by them, i.e., each carrying $n - 4$ calls. Assign $[2n - 7, 2n - 4]$ to cell $(2, 1)$.

Next we delete $[2n - 11, 2n - 8]$ from cells $(1, 1)$ and $(2, 2)$, do the balancing and assign $[2n - 11, 2n - 8]$ to cell $(1, 2)$. Repeatedly doing so, eventually (the last step may delete only two calls) we reach a state consisting of $2n - 2$ distinct indices in cells $(1, 2)$ and $(2, 1)$. Thus a new $(1, 1)$ request must be carried by M_{2n-1}. □

With more involved arguments, necessary and sufficient conditions for $C(n_1, r_1, m, n_2, r_2)$ to be WSNB under P, STU, MI, CS, CD have all been obtained by Chang-Guo-Hwang-Lin and Chang.

Finally, Tsai–Wang–Hwang (2001) proved that for r large, then no algorithm can do better than $2n - 1$. The proof is in the same vein as an argument used by Freeman, Freidman and Pippenger (1986) in proving a multicast rearrangeable result (see Lemma 4.3.4).

Theorem 2.1.12. *For $r \geq (n-1)\binom{2n-2}{n-1} + 1$, $C(n, m, r)$ is WSNB if and only if $m \geq 2n - 1$.*

Proof. It suffices to prove the "only if" part. Suppose $r = (n - 1)\binom{2n-2}{n-1} + 1$ and $m = 2n - 2$. Consider the frame which involves $n - 1$ inputs of every input crossbar and doesn't involve the output crossbar O. Consider the $\binom{2n-2}{n-1}$ distinct $(n - 1)$-subsets of middle crossbars. The $n - 1$ connections of each input crossbar is routed by one such subset. For the given r there must exist a subset Y assigned to a set X of n input crossbars. Add the n new requests $\{(x, o) : x \in X, o \in O\}$ to the frame. Since X is already routed through Y, the new requests cannot use Y anymore. Since the new requests involve the same output crossbar, they must be routed through distinct middle crossbars. Therefore

$$m \geq |Y| + |X| = 2n - 1.$$

□

Note that further increasing r does not lead to a stronger result since O can be connected to at most n input crossbars.

2.2   Constructions of (almost) d-nary Networks

A d-nary nonblocking MIN is usually constructed from a d-nary blocking MIN by adding hardware. There are three general ways of doing this: (i) extend the number of stages, (ii) take parallel copies of ν and identify their input and output switches, (iii) increase the sizes of crossbars(multiple links allowed). We will discuss these alternatives.

Take a binary MIN as our basic network. Can we add enough extra stages to make it nonblocking? The answer is no for SNB. Note that there exist two node (switch)-disjoint paths in any nontrivial network. Consider a state of N connections including these two paths. Delete these two connections and generate two new requests by interchanging the two outputs. The only links available in the network are those just released from the two disjoint paths. Clearly, the two paths cannot interchange their destinations.

For WSNB, Smyth (1990) gave an affirmative answer.

For a binary network, let a path be identified by its input, and every crossbar is labeled by the path (or the two paths) it carries. Smyth proved the following two results.

Lemma 2.2.1. *In a binary network a state s is unsafe if there exist two paths i, i' such that no crossbar is labeled by (i, i').*

Proof. Suppose to the contrary that s is safe under algorithm A. Let path i connect (i, o) and path i' connect (i', o'). Arbitrarily pair up all idle inputs and outputs in s. Since s is safe, these pairs can be connected according to A to a state t. Note that every crossbar carries two paths in t. Delete the two paths i and i'. Then the new request (i, o') cannot be routed since the only links available are those on the (i, o) path and the (i', o') path which do not intersect.   □

Lemma 2.2.1 can be straightfowardly generalized to d-nary networks.

Theorem 2.2.2. *The cost of a WSNB binary multistage N-network ν is at least $4[N(N-1) - \lfloor N^2/4 \rfloor]$.*

Proof. Let $CL(S)$ denote the closure of safe states. If ν is WSNB, $CL(S)$ must contain a maximal state s, a state where all inputs and outputs are busy. Suppose in s there are three inputs i_1, i_2, i_3 where each pair labels only one crossbar. Since ν is multistage, the three crossbars with labels $(i_1, i_2), (i_2, i_3)$ and (i_1, i_3) appear in different stages. Without loss of generality, assume the three pairs appear in the order $(i_1, i_2), (i_2, i_3), (i_1, i_3)$. Switch the connections of i_2 and i_3. Then the only effect on routing is that the i_2-path and the i_3-path switch their subpaths after the crossbar (i_2, i_3). The corresponding changes of labels are; "all labels after (i_2, i_3) should switch i_2 with i_3." Therefore the three labels involving pairs of i_1, i_2, i_3 are now $(i_1, i_2), (i_2, i_3), (i_1, i_2)$. By Lemma 2.2.1, the new state is unsafe since the label (i_1, i_3) is missing.

Therefore for any three inlets, at least one pair should be repeated as a label. Construct a graph with inlets as vertices and unrepeated pairs as edges. Then the graph should be triangle-free. It is well known (Mantel, 1907) such a graph has at most $\lfloor N^2/4 \rfloor$ edges. Therefore at least $2\binom{N}{2} - \lfloor N^2/4 \rfloor$ crossbars are required. Theorem 2.2.2 follows immediately.   □

Halpenny (1990) proved an upper bound of the cost. By connecting idle inputs and outputs arbitrarily, called idle connection, we may assume the network in a maximal state. A maximal state is safe if for any two paths i, j, there exists a crossbar X labeled (i, j) such that interchanging labels i and j from X on is still a safe state. The corresponding algorithm is to route a request (i, k) by interchanging labels i and j from X on if the current state routes j to k. Note that i, k in a request implies that both i and k are idle, in other words, the two connections involving i and k and idle connections. Hence interchanging i and j does not affect any real connection. Thus we may represent a request by the two inputs involved.

Theorem 2.2.3. *There exists a WSNB binary MIN with cost $2N(2^N - N - 1)$.*

Proof. We first construct a binary network H_N which is not in the multistage form. H_N is constructed recursively (H'_{N-1} is a copy of H_{N-1}).

It suffices to prove H_N is WSNB for maximal states.

The safe states consist of all states with at most one "bend" setting of middle switches. Label each crossbar by the paths it carries. We show that a safe state contains all pairs of labels and any interchanging of two paths

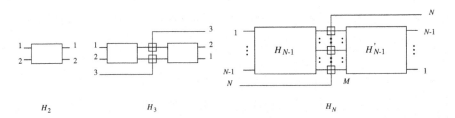

Figure 2.2.1: A recursive construction of H_N.

induces another safe state. The proof is by induction on N. The case of $N = 2$ is trivial.

First consider the case of no "bend" middle switch. Since H_{N-1} is WSNB it contains all labels (i, j) not involving N, and so does H'_{N-1}. The middle stage M contains all labels (i, N). The routing is by interchanging labels in H_{N-1} or M according to whether the request involves N. Note that the new state also has 0 or 1 bend.

Next consider the case of M having exactly one bend switch, say S, with label (k, N). Then H_{N-1} contains all labels (i, j) not involving N, and H'_{N-1} contains all labels (i, j) not involving k. We interchange labels in H_{N-1} or H'_{N-1} accordingly, except when a request involving either input N or output N comes up. Note that the interchange of (k, N) at S can either due to a request (k, N) or a request (N, x) for some output x. In the former case input N can be idle and requests to connect to some output which can be done in H'_{N-1}. In the latter case, output N can be idle and some input i can request to connect to it. This can be done by interchanging label (i, k) in H_{N-1}. Note that the number of bent middle switch is not increased.

By solving

$$C(H_N) = 2C(H_{N-1}) + 4(N - 1),$$

$$C(H_2) = 4,$$

we obtain

$$C(H_N) = 4(2^N - N - 1).$$

Finally, we convert H_N to a binary MIN. Consider the state where every crossbar is set straight. Label a crossbar by the two inlets whose paths go through it. Note that each path travels the crossbars in the direction from left to right and from down to up. Therefore we can partially order the $2^N - N - 1$ crossbars such that the ordering of each path is preserved. Assume N is even.

Add $N/2 - 1$ 2×2 crossbars to each stage and label each crossbar by a pair of inlets arbitrarily except that the set of labels at each stage is the set $\{1, \ldots, N\}$. Insert a link between two crossbars at adjacent stages if and only if they share a label (two links if they share two labels). Note that this construction preserves all paths in H_N. The cost of this network is $2N(2^N - N - 1)$. $\square$

Cantor (1971) proposed the network $[X_{2m}, B_{n-1}, X_{m2}]^*$, where $*$ on X_{2m} denotes that one inlet (outlet) is deleted from each half of the input (output) stage. We will refer to this network as the shaded Cantor network. Thus a $(2n + 1)$-stage shaded Cantor network has $2^n - 2$ inputs and outputs. Before we discuss sufficient conditions for the shaded Cantor network to be SNB, we need the following lemma. A network is said of type $T(p, q)$ if given any state of the network and p idle inputs, then any of the p idle inlets can reach $p + q - 1$ outputs. For example an $a \times b$ crossbar is of types $(p, b - a + 1)$.

Lemma 2.2.4. *Suppose L_{ab} is of type $T(p, q)$, and L_{cd} a SNB network such that $qd \geq a(c - 1)$. Then $L_{ab} \times L_{cd}$ is of type $T(p, qd - a(c - 1))$.*

Proof. Consider a set P of p idle inputs of $L_{ab} \times L_{cd}$. Let $i \in P$ lie in input crossbar A which contains $p' \leq p$ idle inputs. Then i has access to $p' + q - 1$ idle outputs of A. Since L_{cd} is SNB, i has access to $(p' + q - 1)d$ outputs of $L_{ab} \times L_{cd}$. In the worst case, every input of $L_{ab} \times L_{cd}$ not in $P \cup A$ is busy and one output i has access to; i still has access to

$$(p' + q - 1)d - [a(c - 1) - (p - p')]$$

$$= \ p + qd - a(c - 1) + p'(d - 1) - d \geq p + qd - a(c - 1) - 1 \text{ outputs}.$$

$\square$

The following lemma simply states the symmetrical version of Theorem 2.1.1 in a different way.

Lemma 2.2.5. *Suppose L_{ab} is of type $(1, q)$ with $2q > b$ and L is SNB. Then $[L_{ab}, L, L_{ba}]$ is SNB.*

We are now ready to prove the Cantor theorem.

Theorem 2.2.6. $[X_{2m}, B_{n-1}, X_{m2}]^*$ *is SNB if $m \geq 2(n - 1)$.*

Proof. Note that $[X_{2m}, B_{n-1}, X_{m2}]^* = [(X_{2m} \times X_{22}^{n-2})^*, X_{22}, (X_{22}^{n-2} \times X_{m2})^*]$. It is easily seen that $X_{2,m}$ is of type $T(p, m-1)$. Since X_{22} is SNB, $X_{2m} \times X_{22}$ is of type $T(p, 2(m-2))$, $(X_{2m} \times X_{22}) \times X_{22}$ is of type $T(p, 2^2(m-3))$. Using Lemma 2.2.4 recursively, $X_{2m} \times X_{22}^{n-2}$ is of type $T(p, 2^{n-2}(m-n+1))$. Hence $(X_{2m} \times X_{22}^{n-2})^*$ is of type $T(p, 2^{n-2}(m-n+1)+1)$. By Lemma 2.2.5, $[(X_{2m} \times X_{22}^{n-2})^*, X_{22}, (X_{22}^{n-2} \times X_{m2})^*]$ is SNB if $2[2^{n-2}(m-n+1)+1] > 2^{n-2}m$, which is the number of outputs in $(X_{2m} \times X_{22}^{n-2})^*$. Thus $m = 2(n-1)$ suffices. $\square$

Corollary 2.2.7. *The cost of a SNB shaded Cantor network is* $4N(\log_2 N)^2$.

Pippenger (1978) showed that a more careful implementation can reduce the cost to $16N(\log_5 N)^2$.

Cantor also proposed a slightly different version $[X_{1m}, B_n, X_{m1}]$, which is commonly referred to as the Cantor network in the literature.

Corollary 2.2.8. *The Cantor network* $[X_{1m}, B_n, X_{m1}]$ *is SNB if* $m \geq n$.

Proof. Since Theorem 2.2.6 shows that the necessary number of B_{n-1} is even for SNB, one can combine every two B_{n-1} into a B_n (by adding two outer stages). However, we have to add one extra B_n to compensate for the $*$ effect in $X_{2,m}^*$. With some corresponding changes in the numbers of inputs and outputs, the resultant network is $[X_{1,n}, B_n, X_{n,1}]$. $\square$

Compare to the shaded Cantor network, the Cantor network has two extra stages, $2N \log_2 N$ more crosspoints, but also two more inlets and outlets. It is easily shown both have $O(N\log^2 N)$ crosspoints.

Define $\gamma = \lfloor \log_2(8d) \rfloor$. Then $2^\gamma > 4d$, Pippenger (1982) considered the network $[X_{1,2^\gamma}, B_n^d, X_{2^\gamma,1}]$ whose center stage has $2^\gamma(N/d)$ crossbars. Since each input and output can reach at least $2^\gamma(N/d) - N \geq (3/4)2^\gamma(N/d)$ center stage crossbars, the network is nonblocking by Lemma 2.2.5 Lin–Pippenger (1994) noted that each random path from input i or output o to the center stage is blocked with probability at most $1/4$. So a random path from i to o has blocking probability at most a half. Therefore with t processors, a batch of t requests can be routed in $\log_2 t$ expected rounds. The updating of data structure and checking of link occupancy take $O(n)$ time at each round. Therefore the parallel randomized algorithm requires $O((\log N)^2)$ time.

Lin and Pippenger also converted their randomized algorithm to a deterministic one. However, a batch of t requests now takes $O((\log N)^5)$ time to route. They noted that the number of processors required by their algorithm is only the number of requests in a batch, while that required by the routing algorithm of Arora–Leighton–Maggs (see below) is proportional to the cost of the network. Shyy–Lea (1991) generalized the Cantor network by replacing the Beneš network B_n in the middle with $BY_{F-1}^{-1}(k)$ (the Cantor network is the special case $k = n - 1$). Denote such a network by $\log_d(N, k, m)$. For the d-nary version. Fig. 2.2.2 illustrates such a network.

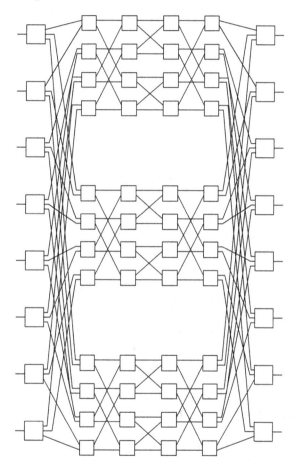

Figure 2.2.2: A $\log_2(8, 1, 3)$ network

Shyy and Lea proved

Theorem 2.2.9. *$Log_2(N, k, m)$ is strictly nonblocking if*

$$m > \begin{cases} k + 3 \cdot 2^{\frac{n-k}{2}-1} - 2, & \text{for } n - k \text{ even,} \\ k + 2^{\frac{n-k+1}{2}} - 2, & \text{for } n - k \text{ odd.} \end{cases}$$

The special case $k = 0$ was first proved by Lea (1991).

Hwang extended Theorem 2.2.9 to the d-nary version with a simplifying proof.

Theorem 2.2.10. $Log_d(N, k, m)$ *is SNB if*

$$
m > \begin{cases}
\frac{2k(d-1)}{d} + (d+1)d^{\frac{n-k}{2}-1} - 2, & \textit{for } n + k \textit{ even,} \\[2mm]
\frac{2k(d-1)}{d} + 2d^{\frac{n-k-1}{2}} - 2, & \textit{for } n + k \textit{ odd.}
\end{cases}
$$

Proof. Consider the channel graph $CG(i, o)$ between an input i and an output o in $BY_{F-1}^{-1}(k)$. A stage-j link may also be seized by a connection (i', o'); where $i' \neq i$ and $o' \neq o$. We call such a connection an *intersecting connection*. To avoid counting twice, we must assign such an intersecting connection either to i' or to o'. It will be made clear later that an intersecting connection has more blocking impact if it is closer to either i or o. Let j be the intersecting stage. To maximize the blocking impact, we assign it to the input side if $j \leq (n + k)/2$, and to the output side otherwise. From the structure of BY^{-1} and the pattern F^{-1}, it is easily verified that $CG(i, o)$ is a symmetric series-parallel channel graph with d-branching at the k outer shells. Let NP_j denote the number of paths at shell j. Then

$$
NP_j = \begin{cases}
d^j, & \text{for } j \leq k, \\[2mm]
d^k, & \text{for } j \geq k.
\end{cases}
$$

Define $NI_j(NO_j)$ as the number of inputs (outputs) which can generate an intersecting connection seizing a shell-j link for the first time. Then because of the buddy property,

$$
NI_j = NO_j = (d-1)d^{j-1} \text{ for } 1 \leq j \leq (n+k)/2 \; \left(\text{except } NO_{(n+k)/2} = 0\right) .
$$

Assuming the worst case that the NI_j and NO_j intersecting connections are all disjoint, then a portion $(NI_j + NO_j)/NP_j$ of the paths in $G(i, o)$ is unavailable to (i, o). Therefore the condition of SNB is

$$m > \sum_{j=1}^{\lfloor (n+k-1)/2 \rfloor} \frac{NI_j + NO_j}{NP_j} = \sum_{j=1}^{k} \frac{2(d-1)d^{j-1}}{d^j}$$

$$+ \sum_{j=k+1}^{\lfloor (n+k-1)/2 \rfloor} \frac{2(d-1)d^{j-1}}{d^k} + \delta(n+k \text{ even})(d-1)\frac{d^{(n+k)/2-1}}{d^k}$$

$$= \frac{2k(d-1)}{d} + \frac{2(d-1)}{d^k} \cdot \frac{(d^{\lfloor (n+k-1)/2 \rfloor} - d^k)}{(d-1)}$$

$$+ \delta(n+k \text{ even})(d-1)d^{(n-k)/2-1}$$

$$= \frac{2k(d-1)}{d} + 2d^{\lfloor (n-k-1)/2 \rfloor} - 2 + \delta(n+k \text{ even})(d-1)d^{(n-k)/2-1},$$

which is the same as the condition given in Theorem 2.2.10. □

Note that NI_j and NO_j are invariant if the basic network of BY^{-1} is replaced by any other buddy network. Therefore Theorem 2.2.10 holds for other buddy networks and other extra-stage addition patterns if their channel graphs are isomorphic to $BY_{F-1}^{-1}(k)$.

The channel graph in Theorem 2.2.10 has three special properties; symmetric, series-parallel and double-tree, the last means both the left and the right sides are trees with their leaves identified or connected one-to-one by edges. The first property affects only the notion of shells, which is non-essential to the method, since we count blocking on the left and right sides independently any way. But the method lives and dies with the second and third properties. This is because in a series-parallel double tree, a busy link in shell i automatically blocks all links in shell $i+1$ connected to it, thus we only need to know at which shell a blocking path first enters to assess its blocking impact.

We comment that NP_j is maximized under BY_{F-1}^{-1} for every j. Therefore BY_{F-1}^{-1} is indeed the best choice among all extra-stage self-routing network with a series-parallel double-tree channel graph.

Corollary 2.2.11. $Log_d(N, n-1, m)$ is SNB if $m > 2(n-1)(d-1)/d$.

Note that $\log_d(N, n-1, m)$ is the d-nary Cantor network. Hence Corollary 2.2.11 extends Corollary 2.2.8 from $d = 2$ to general d.

It is easily verified that under the condition of Theorem 2.2.10, the cost is minimized by setting $k = n - 1$. More specifically, the cost reduces from $O(N^{\frac{3}{2}} \log N)$ for $k = 0$ to $O(N \log^2 N)$ for $k = n - 1$.

BY^{-1} without its inputs and outputs is called a *butterfly*. Figure 2.2.3 illustrates a 3-stage butterfly. Note that stage-j links can be decomposed into 2^j components. Let B and B' be two s-stage butterflies except that a permutation $\pi_j = (0, 1, \ldots, 2^{s-j} - 1) \rightarrow (0, 1, \ldots, 2^{s-j} - 1) \pmod{2^{s-j}}$ is associated with stage j of B' for $1 \leq j \leq s$. Upfal (1989) defined a 2-butterfly as a superimposition of B and B' by identifying the crossbars at every stage. Figure 2.2.3 illustrates a 2-butterfly.

$$B \qquad\qquad\qquad B' \qquad\qquad\qquad B \otimes B'$$

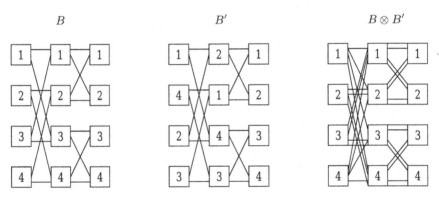

Figure 2.2.3: A 2-butterfly

Similarly, a d-butterfly can be defined by superimposing d butterflies each carrying a set of permutations (one of them is a set of identity permutations). Each node is of size $2d \times 2d$ except only one input (output) is assigned to each input (output) node.

A component of stage-j links is called a *splitter* since half of its links go to nodes belonging to one component of stage-$(j + 1)$ links, and the other half to another. We call these two types the *up succeeding nodes* and the *down succeeding nodes*. By construction, all splitters of a given stage are isomorphic. An M-splitter, where M is the size of preceding nodes, is said to have the (α, β) expansion property if every set of $m \leq \alpha M$ *preceding nodes* is adjacent to at least $\beta m, \beta > 1$, of up succeeding nodes and down succeeding nodes each. A random permutation has a good probability of landing the (α, β) expansion property for any d, α, β satisfying $2\alpha\beta < 1$ and $2d > 2\beta + 1 + (2\beta + 1 + \ln 2\beta)/\ln(1/2\alpha\beta)$. Explicit permutations with the expansion property are also known but not as good.

A d-Beneš network consists of the concatenation of the mirror image of a d-butterfly with the d-butterfly. A d-Beneš network is also called a *multi-Beneš network* when d is not specified. The counterpart of a splitter is called a *merger*. The (α, β) expansion property of a merger is similarly defined as for a splitter except that $2\beta m$ succeeding nodes do not have to split half and half into "up" and "down".

Arora–Leighton–Maggs (1990) proposed the multi-Beneš network as a SNB with a $O(\log N)$ time routing algorithm. The analysis on nonblockingness and the routing will be given together. Because in the construction of a d-Beneš network, the mirror image precedes the d-butterfly, the first half of the network consists of mergers, and the second half splitters.

In order for the multi-Beneš network to be nonblocking, we must reduce the number of inputs and outputs while keeping the network itself intact. Suppose we keep only every L^{th} input (output). Then the first and last $\ell = \log_2 L$ stages can be eliminated with each input (output) occupying L designated crossbars at stage ℓ(stage $2n + 1 - \ell$).

A node is called *busy* if it carries a connection, and is called *blocked* if all its up succeeding nodes or all its down succeeding nodes or all its $2d$ succeeding nodes are either busy or blocked. A node is *nonworking* if it is busy or blocked, and is *working* otherwise.

Lemma 2.2.12. *For $L > 1/2\alpha(\beta - 1)$ at most an α fraction of the preceding nodes of a splitter are blocked because of nonworking up succeeding nodes, and at most an α fraction are blocked because of nonworking down succeeding nodes.*

Proof. The proof is by induction on the stage number k, Lemma 2.2.12 is trivially true for $k = 2n - \ell$ since at most one up (down) succeeding node is nonworking (busy) in every splitter. Consider the general k case. Suppose the preceding nodes of an M-splitter contain more than αM nodes that are blocked because of the up succeeding nodes. By the (α, β) expansion property, the αM nodes are adjacent to at least $\beta \alpha M$ up succeeding nodes, and all of them are blocked. But only $M/2L$ outputs have access to the $M/2$ up succeeding nodes and make them busy. Furthermore, by induction, at most $2\alpha(M/2) = \alpha M$ up succeeding nodes are blocked. Therefore $\beta \alpha M \leq M/2L + \alpha M$, or $L \leq 1/2\alpha(\beta - 1)$, contradicting our assumption. □

Corollary 2.2.13. *For $L > 1/2\alpha(\beta - 1)$ at most a 2α fraction of the preceding nodes in any splitter are blocked.*

Similarly, we can prove

Lemma 2.2.14. *For $L > 1/2\alpha(\beta-1)$, at most a 2α fraction of the up preceding nodes and a 2α fraction of the down preceding nodes in any merger are blocked.*

Suppose that each node has processing ability. By working backwards from stages $2n + 1 - \ell$ to stage ℓ, all blocked nodes can be identified in $O(n)$ time by a fault propagation process. Since at most a $2\alpha < 1$ fraction of the preceding nodes of any merger are blocked, an idle input has access to a working stage-ℓ node u. Since u is working, by definition, u has access to a stage-$(n + 1)$ working node w. Similarly, every working second half node has access to a working up-neighbor and a working down-neighbor. Hence w has access to every output. A greedy routing algorithm can be easily implemented in $O(dn)$ time, or $O(n)$ time if d is fixed.

Arora–Leighton–Maggs also showed that $O(n)$ time also suffices to route many paths at once.

Superposing d copies should always be better than a vertical layout due to shared resource. However the ALM network is not better than Theorem 2.2.9 for n small. This is because the expander property does not make use of duplicate edges.

2.3   Multistage Networks

Since Shannon (1950) proved that $\Omega(N \log N)$ is a lower bound of the cost of a rearrangeable network, it must remain one for both SNB network and a WSNB network. On the other hand, Bassalygo–Pinsker (1974) first proved the existence of an $O(N \log N)$ SNB network by using expanders as components. The general idea is to use the expanders recursively until each input is guaranteed to reach the majority of outputs. The concatenation of this network and its inverse then yields a SNB network. Explicit constructions of these expanders were later given by Margulis (1973), Gabber-Galil (1981), Jimbo-Maruoka (1985) and Lubotzky-Phillips-Sarnak (1986).

For fixed number s of stages Friedman (1988) derived a lower bound of cost by the following argument. He considered the graph version of a MIN. Let $d(v)$ denote the indegree of a node v, and $d'(v)$ its outdegree.

Lemma 2.3.1. *Consider the channel graph $CG(i, o)$ of the graph of a SNB s-stage network G. Let A_{io} denote the set of stage-2 nodes and B_{io} the set of* *stage-$(s-1)$ nodes in $CG(i,o)$. Then*

$$\sum_{v \in A_{io}} \frac{1}{d(v)} + \sum_{u \in B_{io}} \frac{1}{d'(u)} > 1 \,.$$

Proof. Without loss of generality, assume $A_{io} = \{v_1, \ldots, v_a\}$ and $B_{io} = \{u_1, \ldots, u_b\}$ such that

$$d(v_1) \le d(v_2) \le \cdots \le d(v_a) \,,$$

$$d'(u_1) \le d'(u_2) \le \cdots \le d'(u_b) \,.$$

Suppose that $d(v_j) \ge j+1$ for all $1 \le j \le a$. Then there exist distinct inputs $i_1, \ldots, i_a$ other than i such that i_j is adjacent to v_j. Similarly, if $d'(u_k) \ge k+1$ for all $1 \le k \le b$, then there exist distinct outputs $o_1, \ldots, o_b$ other than o such that o_k is adjacent to u_k. Let M denote a maximal set of node-disjoint paths between A_{io} and B_{io} and let M' denote the set of corresponding paths between $\{i_1, \ldots, i_a\}$ and $\{o_1, \ldots, o_b\}$. Then in a state containing M' but i and o are idle, the request (i,o) cannot be connected, contradicting the assumption that G is SNB.

Therefore either $d(v_j) \le j$ for some $1 \le j \le a$ or $d'(u_k) \le k$ for some $1 \le k \le b$. Assume the former. Then

$$\sum_{k=1}^{j} \frac{1}{d(v_k)} \ge \sum_{k=1}^{j} \frac{1}{j} = 1 \,.$$

$\square$

Lemma 2.3.2. *Let $G(V,E)$ be a SNB s-stage network such that $d(v) \le \Delta$ and $d'(v) \le \Delta$ for all $v \in V$. Then*

$$N < 2\Delta^{s-1} \,.$$

In particular,

$$N^2 < |E| \qquad for \ s = 2.$$

Proof. Let A_o denote the set of stage-2 nodes which has a path to o, and let B_i denote the set of stage-$(s-1)$ nodes which i has a path to. Then $A_{io} \subseteq A_o$, $B_{io} \subseteq B_i$.

Summing the inequality in Lemma 2.3.1 over all i and o, we obtain

$$N^2 < \sum_{i \in I} \sum_{o \in O} \left(\sum_{v \in A_{io}} \frac{1}{d(v)} + \sum_{u \in B_{io}} \frac{1}{d'(u)} \right)$$

$$= \sum_{o \in O} \sum_{v \in A_o} (\text{number of inputs adjacent to } v) \, \frac{1}{d(v)}$$

$$+ \sum_{i \in I} \sum_{u \in B_i} (\text{number of outputs adjacent to } u) \, \frac{1}{d'(u)}$$

$$= \sum_{o \in O} \sum_{v \in A_o} 1 + \sum_{i \in I} \sum_{u \in B_i} 1 \le \sum_{o \in O} (\Delta)^{s-1}$$

$$+ \sum_{i \in I} (\Delta)^{s-1} = 2N \, (\Delta)^{s-1} \, .$$

Note that for $s = 2$

$$\sum_{o \in O} \sum_{v \in A_o} 1 + \sum_{i \in I} \sum_{u \in B_i} 1 = |E|.$$

Lemma 2.3.2 follows immediately.   □

Theorem 2.3.3. *Let G be a SNB s-stage network for $s \ge 3$. Then the cost of G is*

$$|E| \ge (\frac{N}{4})^{1 + \frac{1}{s-1}} \, .$$

Proof. Let V' be the set of nodes such that $v \in V'$ implies either $d(v) \ge 4|E|/N$ or $d'(v) \ge 4|E|/N$. Since there are at most $N/4$ nodes with $d(v) \ge 4|E|/N$, and at most $N/4$ nodes with $d'(v) \ge 4|E|/N$, $|V'| \le N/2$. Let M be a maximal set of node-disjoint paths from inputs to outputs such that each path hits a member of V'. Define $V'' = V - V' - (\text{nodes in } M)$. Then the subnetwork G'' induced by V'' is also SNB with at least $N/2$ inputs and $N/2$ outputs. Apply Lemma 2.3.2 to G'' by setting $\Delta = 4|E|/n$. Then

$$\frac{N}{2} < 2(\frac{4|E|}{N})^{s-1}$$

or

$$|E| > (\frac{N}{4})^{1 + \frac{1}{s-1}} \, .$$

□

Note that Friedman's original version has the coefficient 4 times smaller due to a weaker Lemma 2.3.2.

Clos proposed a multistage extension of the 3-stage Clos network by iteratively replacing each crossbar in the middle stage with a SNB 3-stage Clos network. Note that the only property of crossbar used in the proof of Theorem 2.1.1 is that it is internally SNB. Therefore the replacement of crossbars by SNB subnetworks does not affect the SNB property of the network. This way we grow from 3 stages to 5 stages to 7 stages. Clos showed that the cost of a $(2k + 1)$-stage network is bounded by $(5 \cdot 2^k - 4)N^{1+\frac{1}{k+1}}$. Later, Pippenger (1978) improved the constant to $2(k + 1)\left(\frac{2^{\binom{k+2}{2}}}{4}\right)^{\frac{1}{k+1}}$. Note that the complexity is $O(N^{1+\frac{2}{s+1}})$.

Kharkevich (1957) and Cantor (1971) independently proposed a more general multistage extension which allows the replacement of crossbars by SNB 3-stage Clos networks in any stage. Let $C_{\text{SNB}}(N, M)$ denote the minimum cost of a SNB (N, M)-network.

Theorem 2.3.4. $C_{\text{SNB}}(N, N) \leq O\left(N(\log N)^{\log_2 5}\right).$

Proof. Consider the network $\nu = [L_{a,2a}, L_{a,2a}, L_{2a,a}]$ which has a^2 inputs and $2a^2$ outputs, where L is SNB. By Theorem 2.1.1 and our remarks on replacement, ν is SNB. Hence

$$C_{\text{SNB}}(a^2, 2a^2) \leq 3aC_{\text{SNB}}(a, 2a) + 2aC_{\text{SNB}}(a, 2a) = 5aC_{\text{SNB}}(a, 2a).$$

Solving this recursive inequality, we obtain

$$C_{\text{SNB}}(a, 2a) \leq O\left(a(\log a)^{\log_2 5}\right).$$

Hence

$$C_{\text{SNB}}(N, N) \leq C_{\text{SNB}}(N, 2N) \leq O\left(N(\log N)^{\log_2 5}\right).$$

$\square$

Next we introduce the *extended generalized shuffle networks* (EGSN) proposed by Richards and Hwang (1999). An EGSN is a MIN where the interconnection pattern between two adjacent stages is the shuffle pattern. More specifically, let v be an s-stage EGSN where stage i consists of $r_i X_{n_i,m_i}$'s, $1 \leq i \leq s$, satisfying $r_i m_i = r_{i+1} n_{i+1}$ for $1 \leq i \leq s - 1$. Consider the interconnection between stages i and $i + 1$. In the shuffle pattern, the m_i links from crossbar x_{ij} are adjacent to crossbars $x_{i+1,jm_i}, x_{i+1,jm_i+1}, \ldots, x_{i+1,(j+1)m_i-1}$, when the second subscript is in modulo r_{i+1} addition. Figure 2.3.1 shows a 6-stage EGSN with $r_1 = 8$, $n_1 = 2$, $m_1 = 2$, $r_2 = 4$, $n_2 = 4$, $m_2 = 3$, $r_3 = 6$, $n_3 = 2$, $m_3 = 3$, $r_4 = 9$, $n_4 = 2$, $m_4 = 2$, $r_5 = 6$, $n_5 = 3$, $m_5 = 4$, $r_6 = 8$, $n_6 = 3$, $m_6 = 2$, which can also be represented by the array: $2\boxed{8}2\ 4\boxed{4}3\ 2\boxed{6}3\ 2\boxed{9}2\ 3\boxed{6}4\ 3\boxed{8}2$.

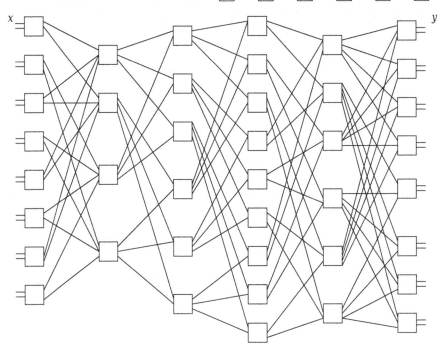

Figure 2.3.1: An EGSN

Define $N_{ij} = \prod_{k=i}^{j} n_k$, $M_{ij} = \prod_{k=i}^{j} m_k$, $N = n_1 r_1$, and $M = m_s r_s$.

Lemma 2.3.5. *The number of paths NP_{ij} from a stage-i crossbar u to a stage-j crossbar v is either $\lceil M_{i,j-1}/r_j \rceil = \lceil N_{i+1,j}/r_i \rceil$ or $\lfloor M_{i,j-1}/r_j \rfloor = \lfloor N_{i+1,j}/r_i \rfloor$.*

Proof. The shuffle pattern guarantees that the paths from u, when properly ordered, hits consecutive crossbars at stage k for all $k = i+1, \ldots, j$ (if the number of paths at stage k exceeds r_k, then the consecutive hitting may run through the set of r_k crossbars more than once). In particular, the $M_{i,j-1}$ paths from x hits the r_j stage-j crossbars consecutively, hence evenly, i.e., the number of paths hitting two stage-j crossbars can differ by at most one. It follows that the number of paths hitting v is either $\lceil M_{i,j-1}/r_j \rceil$ or $\lfloor M_{i,j-1}/r_j \rfloor$. Counting the paths from v to u, we obtain $\lceil N_{i+1,j}/r_i \rceil$ or $\lfloor N_{i+1,j}/r_i \rfloor$. $\square$

Corollary 2.3.6. NP_{1j} *is either* $\lceil N_{1j}/N \rceil$ *or* $\lfloor N_{1j}/N \rfloor$; NP_{is} *is either* $\lceil M_{is}/M \rceil$ *or* $\lfloor M_{is}/M \rfloor$.

Proof. $NP_{1j} = \lceil N_{2j}/r_1 \rceil$ implies $NP_{1j} = \lceil N_{1j}/r_1 n_1 \rceil = \lceil N_{1j}/N \rceil$. The other case is similar. □

Corollary 2.3.7. *The number of paths in a channel graph is either* $\lceil M_{1s}/M \rceil = \lceil N_{1s}/N \rceil$ *or* $\lfloor M_{1s}/M \rfloor = \lfloor N_{1s}/N \rfloor$.

The general approach here to derive a sufficient condition for strict non-blockingness is also through the study of channel graph as Shyy–Lea did in Theorem 2.2.9. However the Channel graph here is no longer a series-parallel double-tree.

Richards–Hwang (1999) devised a new method to deal with a general channel graph. Instead of treating each link in the channel graph as an independent object to be seized by other connections, they consider the possibility that a connection intersects more than one link in the channel graph. However, whenever that happens, those intersected links must be consecutive. This is because if a connection intersects two nonconsecutive links e and e' in $G(x,y)$, then the links between e and e' in the connection must also be in $G(x,y)$. Let $C(i,j)$ denote a connection which enters $G(x,y)$ at stage i and exists at stage j. We determine $B[C(i,j)]$, the maximum number of paths in $G(x,y)$ blocked by $C(i,j)$; or equivalently, the maximum number of paths in $G(x,y)$ using a link of $C(i,j)$ from stage i to stage j.

Let t denote the largest k such that $N_{1k} < N$, and u the smallest k such that $M_{ks} < M$. Namely, for $k > t$, each stage-k crossbar has access to every input; or equivalently, each input has access to every stage-k crossbar. Similarly, for $k < u$, every output and every stage-k crossbar have access to each other. Then $C(i,j)$ must satisfy $i \le t + 1$ and $j \ge u - 1$. To see this suppose $i > t + 1$. Let u and v be the stage-$(i-1)$ and stage-i node on $C(i,j)$. Since $i - 1 \ge t + 1$, both u and v are in $G(x,y)$. Hence $C(i,j)$ enters $G(x,y)$ at stage $i - 1$, contradicting the definition of $C(i,j)$. The proof of $j \ge u - 1$ is similar.

Lemma 2.3.8.

$$B[C(i,j)] \le \left\lceil \frac{M_{i+1,s}}{M} \right\rceil + \left\lceil \frac{N_{i,j-1}}{N} \right\rceil + c(t,u)$$

$$c(t,u) = \begin{cases} \sum_{k=t+1}^{u-2} \left\lceil \frac{N_{1k}}{N} \right\rceil \left(\left\lceil \frac{M_{k+1,s}}{M} \right\rceil - \left\lceil \frac{M_{k+2,s}}{M} \right\rceil \right) - \left\lceil \frac{M_{t+2,s}}{M} \right\rceil, & \text{if } u - 2 \ge t + 1, \\ -1, & \text{if } u - 2 \le t. \end{cases}$$

Proof. Let λ_k, $i \leq k \leq j-1$, denote a stage-k link on $C(i,j)$, and let u and v denote the stage-i and stage-$(i+1)$ node of λ_i. There are NP_{1i} paths from x to u, and $NP_{i+1,s}$ paths from v to y. Therefore, the number of paths in $G(x,y)$ containing λ_i is $NP_{1i} \cdot NP_{i+1,s}$. Similarly, the number of paths in $G(x,y)$ containing λ_{i+1} is $NP_{1,i+1} \cdot NP_{i+2,s}$. However, those paths containing both λ_i and λ_{i+1} are counted twice and need to be subtracted; and there are $NP_{1i} \times NP_{i+2,s}$ of them. Using this argument repeatedly, then each λ_k, $i+1 \leq k \leq j-1$, blocks an extra $NP_{1k} \cdot NP_{k+1,s} - NP_{1,k-1} \cdot NP_{k+1,s}$ paths. Adding up,

$$B[C(i,j)] = \sum_{k=i}^{j-1} NP_{1k} \cdot NP_{k+1,s} - \sum_{k=i}^{j-2} NP_{1k} \cdot NP_{k+2,s}$$

$$= \begin{cases} \sum_{k=i}^{j-1} NP_{1k} \left(NP_{k+1,s} - NP_{k+2,s} \right) + NP_{1,j-1} \cdot NP_{j+1,s} \\[2ex] \sum_{k=i+1}^{j-1} \left(NP_{1k} - NP_{1,k-1} \right) NP_{k+1,s} + NP_{1,i} \cdot NP_{i+1,s}. \end{cases}$$

Since $NP_{k+1,s} \geq NP_{k+2,s}$ and $NP_{j+1,s} \geq 0$, in the top equation, $B[C(i,j)]$ is upper-bounded by setting NP_{1k} to their upper bounds. From the bottom equation, $NP_{k+1,s}$ should be similarly set. By Corollary 2.3.6 and noting $i-1 \leq t, j+1 \geq u$, $\lceil N_{1k}/N \rceil = 1$ for $k \leq t$ and $\lceil M_{ks}/M \rceil = 1$ for $k \geq u$.

$$B[C(i,j)] \leq \sum_{k=i}^{j-1} \left\lceil \frac{N_{1k}}{N} \right\rceil \left(\left\lceil \frac{M_{k+1,s}}{M} \right\rceil - \left\lceil \frac{M_{k+2,s}}{M} \right\rceil \right) + \left\lceil \frac{N_{1,j-1}}{N} \right\rceil \left\lceil \frac{M_{j+1,s}}{M} \right\rceil.$$

If $u-2 \geq t+1$, then

$$B[C(i,j)] \leq \sum_{k=i-1}^{j-1} \left\lceil \frac{N_{1k}}{N} \right\rceil \left(\left\lceil \frac{M_{k+1,s}}{M} \right\rceil - \left\lceil \frac{M_{k+2,s}}{M} \right\rceil \right) - \left\lceil \frac{M_{is}}{M} \right\rceil$$

$$+ \left\lceil \frac{M_{i+1,s}}{M} \right\rceil + \left\lceil \frac{N_{1,j-1}}{N} \right\rceil$$

$$= \sum_{k=i-1}^{t} \left(\left\lceil \frac{M_{k+1,s}}{M} \right\rceil - \left\lceil \frac{M_{k+2,s}}{M} \right\rceil \right) + \sum_{k=t+1}^{u-2} \left\lceil \frac{N_{1k}}{N} \right\rceil$$

$$\cdot \left(\left\lceil \frac{M_{k+1,s}}{M} \right\rceil - \left\lceil \frac{M_{k+2,s}}{M} \right\rceil \right) - \left\lceil \frac{M_{is}}{M} \right\rceil + \left\lceil \frac{M_{i+1,s}}{M} \right\rceil + \left\lceil \frac{N_{1,j-1}}{N} \right\rceil$$

$$= \left\lceil \frac{M_{i+1,s}}{M} \right\rceil + \left\lceil \frac{N_{1,j-1}}{N} \right\rceil + \sum_{k=t+1}^{u-2} \left\lceil \frac{N_{1k}}{N} \right\rceil$$

$$\cdot \left(\left\lceil \frac{M_{k+1,s}}{M} \right\rceil - \left\lceil \frac{M_{k+2,s}}{M} \right\rceil \right) - \left\lceil \frac{M_{t+2,s}}{M} \right\rceil.$$

If $u - 2 \leq t$, then

$$B[C(i,j)] \leq \sum_{k=i}^{u-2} \left(\left\lceil \frac{M_{k+1,s}}{M} \right\rceil - \left\lceil \frac{M_{k+2,s}}{M} \right\rceil \right) + \left\lceil \frac{N_{1,j-1}}{N} \right\rceil \left\lceil \frac{M_{j+1,s}}{M} \right\rceil$$

$$= \left\lceil \frac{M_{i+1,s}}{M} \right\rceil - 1 + \left\lceil \frac{N_{1,j-1}}{N} \right\rceil.$$

$\square$

Note that $B[C(i,j)]$ consists of one term depending on i only, one term depending on j only, and the other terms independent of i and j.

The next logical step would be to determine for each pair (i,j) how many $C(i,j)$ are there. But due to Lemma 2.3.8 we need only to determine how many connections enter at stage i, and how many exit at stage j. First we give an upper bound of the total number ω of intersecting connections. This number is greatly reduced by the buddy property which EGSN possesses when certain divisibility conditions are met.

Lemma 2.3.9. *EGSN has the (i,j) buddy property if $M_{i,j-1}$ divides r_j or vice versa.*

Proof. The paths from consecutive stage-i crossbars hit stage-j crossbars consecutively, it is clear that

$$V_j(u) = V_j(v) \quad \text{if } u \equiv v \pmod{q},$$

and

$$V_j(u) \cap V_j(v) = \emptyset \quad \text{otherwise}.$$

$\square$

Corollary 2.3.10. *EGSN has the $(1, k-1)$ buddy property if one of $(N_{1,k-1}, N)$ divides the other, it has the $(k+1, s)$ buddy property if one of $(M_{k+1,s}, M)$ divides the other.*

Proof. One of $(M_{1,k-2}, r_{k-1})$ divides the other is equivalent to one of $(N_{2,k-1}, r_1)$ divides the other is equivalent to one of $(N_{1,k-1}, N)$ divides the other. The proof of the other part is similar. $\square$

We will refer to the conditions in Corollary 2.3.10 as the *divisibility condition at stage k*.

Define $\omega(k) = N_{1,k-1} + M_{k+1,s} - 2$, if $N_{1,k-1}$ divides N and $M_{k+1,s}$ divides M, i.e., $\omega(k)$ is defined only when the divisibility condition, or the buddy property, is satisfied at stage k.

Lemma 2.3.11. $\omega(k)$ *is an upper bound of the number of intersecting connections.*

Proof. An intersecting connection can be either counted at its entry stage or its exit stage. Let stage k be the dividing stage, then each intersecting connection must either enter before or exit after stage k and we count the sum of these two sets of intersecting connections (note that an intersecting connection both entering before and exiting after stage k is still counted twice). By the $(1, k-1)$ buddy property, the set of input switches having access to $V_{k-1}(x)$ is the same set having access to just one of them. Furthermore, this set also contains the set of input switches having access to $V_i(x)$ for all $1 \leq i < k - 1$. Let $u \in V_{k-1}(x)$. Then u has access to $N_{1,k-1}$ inputs including x. It follows that $N_{1,k-1} - 1$ inputs can generate an intersecting connection entering at or before stage $k - 1$. Similarly, $M_{k+1,s} - 1$ outputs can generate an intersecting connection exiting at r after stage $k + 1$. Thus

$$N_{1,k-1} - 1 + M_{k+1,s} - 1 = N_{1,k-1} + M_{k+1,s} - 2$$

is an upper bound. ⊏

Define

$$\omega = \min \{\omega(k) \text{ is defined} : k = 2, \ldots, s - 1, N - 1, M - 1\}.$$

The next step is to assign these ω intersecting connections to the entry stages and the exit stages to maximize the number B of paths blocked in $G(x, y)$. Since $\lceil M_{i+1,s}/M \rceil$ is decreasing in i, and $\lceil N_{1,j-1}/N \rceil$ is increasing in j, the smaller the entry stage and the larger the exit stage yield the more blocked paths. Noting that at most $\min\{N_{1i} - 1, \omega\}$ intersecting connections can enter at stage i or before and at most $\min\{M_{js} - 1, \omega\}$ intersecting connections can exit at stage j or after, we have

$$B = \sum_{i=1}^{s} [\min\{N_{1i} - 1, \omega\} - \min\{N_{1,i-1} - 1, \omega\}] \lceil M_{i+1,s}/M \rceil$$

$$+ \sum_{j=1}^{s} [\min\{M_{js} - 1, \omega\} - \min\{M_{j+1,s} - 1, \omega\}] \lceil N_{1,j-1}/N \rceil$$

$$+ wc(t, u),$$

where $N_{10} \equiv M_{s+1,s} \equiv 0$. Summarizing, we have

Theorem 2.3.12. *An EGSN is SNB if* $B < \lfloor M_{1s}/M \rfloor$.

Example. Consider the EGSN: $3\,\boxed{36}\,22\ \ 6\,\boxed{132}\,2\ \ 6\,\boxed{44}\,7\ \ 4\,\boxed{77}\,4\ \ 11\,\boxed{28}\,4.$
Then $t = u = 3$, $N = 108$, $M = 112$ and $M_{1s}/M = 44$.

$$\omega = \min\{3{+}112{-}2,\ 18{+}16{-}2,\ 108{+}4{-}2,\ 108{-}1,\ 112{-}1\} = 32.$$
$$B = (2{-}0)2 + (17{-}2)1 + (32{-}17)1 + (3{-}0)4 + (15{-}3)1 + (32{-}15)1 - 32$$
$$= 43 < 44.$$

Hence the EGSN is SNB.

2.4 Standard Path Networks

Only binary networks are considered in this section. In a standard path network (SPN), the connection from i to o is fixed for any i and o. Hence each link ℓ can be labeled by the set $P(\ell)$ of standard paths it carries. Consequently, each crossbar can be classified according to the "intersection" relation between its two $P(\ell)$ sets from its inlets and the two $P(\ell)$ sets from its outlets. Halpenny–Smyth (1992) and Hollmann–van Lint (1992) independently showed that only a few types of "standard-path crossbars" exist, and determined the minimum number of crossbars required by a SPN. Our presentation follows closer to the Halpenny–Smyth approach.

Let $i : j$ denote a standard path from input i to output j. i is called the *first label* and j the *second label*. Let $i : \bar{m}$, called a *vertical label*, denote the set of standard paths from i to an m-set of outputs. The label $\bar{m} : j$, called a *horizontal label*, is similarly defined. Note that the label $i : j$ is a special case of $i : \bar{m}$ or $\bar{m} : j$.

Lemma 2.4.1. *Every link label is either $i : \bar{m}$ or $\bar{m} : j$.*

Proof. A link labeled by both (i_1, j_1) and (i_2, j_2) with $i_1 \neq i_2$ and $j_1 \neq j_2$ would have one connection blocked while carrying the other. $\qquad\square$

Lemma 2.4.2. *Suppose S is a subset of the plane which can be partitioned in two ways as a disjoint union of two sets $S = S_{11} \cup S_{12} = S_{21} \cup S_{22}$, where S_{ij} lies on a horizontal or a vertical line. Then either*

(i) *S lies on a horizontal or vertical line, or*

(ii) *S is the union of a horizontal line and a vertical line meeting at a point p (see Fig. 2.4.1).*

(iii) *S consists of the four corners of a rectangle.*

Figure 2.4.1: Case (ii).

Proof. If (i) is not true, then S consists of two lines. Suppose one is horizontal and the other vertical. Then (ii) is the only possibility, with p belonging to the horizontal line in one partition, and to the vertical line in the other. Suppose both are horizontal or both are vertical. Then (iii) is the only possibility, with one partition consisting of the two vertical pairs of points, and the other the two horizontal pairs.   □

Theorem 2.4.3. *There exist ten types of standard-path crossbars. Five of them are shown in Fig. 2.4.2; the other five are obtained by transposing the vertical and horizontal labels.*

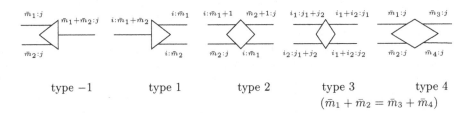

Figure 2.4.2: Types of standard-path crossbars.
(Their transposes are called types $-1^*, 1^*, 2^*, 3^*$ and 4^*.)

Proof. Interpret the label $i : j$ as a point with coordinates (i, j) in the plane. Then $\bar{m} : j$ is a horizontal line and $i : \bar{m}$ a vertical line. Consider the set of labels on the inlet side of a standard-path crossbar. It represents a partition of the set of points which is the union of the set (if the set consists of one label, then it is a trivial partition). Similarly, the set of labels on the outlet side of that crossbar represents another partition of the same set of points. By Lemma 2.4.2, there are only three cases. Types -1, 1 and 4 correspond to case (i), type 2 corresponds to case (ii), and type 3 corresponds to case (iii). By interchanging horizontal lines with vertical lines, the other five types are obtained.   □

Consider a SPN. Let N_i denote the number of standard-path crossbars of type i.

Lemma 2.4.4. (i) $(N_{-1} - N^*_{-1} + N_1 - N^*_1) + 2(N_2 - N^*_2) + 4(N_3 - N^*_3)$
$= 2N(N - 1)$,

(ii) $2(N_{-1} + N_1 + N_2) + (N_4 + N^*_4) \geq 2N(N - 2)$.

Proof. For a link ℓ let $\#_1(\ell)$ denote its number of first labels, and $\#_2(\ell)$ its number of second labels. For a crossbar X, let $L(X)$ denote its inlinks and $R(X)$ its outlinks. Define

$$\#_1(L(X)) = \sum_{\ell \in L(X)} \#_1(\ell),$$

$$\#'_1(L(X)) = \sum_{\ell \in L(X)} \max\{\#_1(\ell) - 2, 0\}$$

and define $\#_1(R(X))$, $\#_2(L(X))$, $\#_2(R(X))$, $\#'_1(R(X))$, $\#'_2(L(X))$ and $\#'_2(R(X))$ similarly. Also define

$$\text{split}(X) = \#_1(R(X)) - \#_1(L(X)) + \#_2(L(X)) - \#_2(R(X)),$$

$$\text{rate}(X) = \max\{0, \#'_1(R(X)) - \#'_1(L(X)) + \#'_2(L(X)) - \#'_2(R(X))\}.$$

Split(X) and rate(X) for various types of X are shown below:

Types	-1	1	2	3	4	-1^*	1^*	2^*	3^*	4^*
split	1	1	2	4	0	-1	-1	-2	-4	0
rate	$0,1,2$	$0,1,2$	$0,1,2$	0	$0,1$	0	0	0	0	$0,1$

Whether the rates of types -1, 1 and 2 are 0, 1 or 2 depends on how many 1 and 2 are there for $\bar{m}_1$ and $\bar{m}_2$. For example, if $\bar{m}_1 = 1$ and $\bar{m}_2 \geq 2$, then the rate for type -1 is

$$\max\{1+\bar{m}_2-2, 0\}-\max\{1-2, 0\}-\max\{\bar{m}_2-2, 0\} = (\bar{m}_2-1)-0-(\bar{m}_2-2) = 1.$$

Input i starts with the label $i : \bar{N}$ and ends with the set $\{i : j, 1 \leq j \leq N\}$ by a sequence of splits, while every split contributes 1 to $\#_1(R(X))-\#_1(L(X))$. Hence i contributes $N - 1$ to the split-value, and the N inputs contribute a total of $N(N - 1)$. (i) follows by also considering the splitting of $\bar{N} : j$.

Similarly, the splittings from $i : \bar{N}$ to $\{i : j, 1 \leq j \leq N\}$ cause a reduction of the $\#'_1$-value from the initial $N - 2$ to 0. Hence the N inputs contribute a total of $N(N - 2)$ to the rate-value. (ii) follows by considering the splitting of $\bar{N} : j$. □

Theorem 2.4.5. *A binary SPN has at least $N^2 - \lfloor 3N/2 \rfloor$ crossbars.*

Proof. $2(N_{-1} + N^*_{-1} + N_1 + N^*_1 + N_2 + N^*_2 + N_3 + N^*_3 + N_4 + N^*_4)$

$$
\begin{aligned}
&= \frac{1}{2} \big\{ \big[(N_{-1} - N^*_{-1} + N_1 + N^*_1) + 2(N_2 - N^*_2) + 4(N_3 - N^*_3) \big] \\
&\quad + \big[2(N_{-1} + N_1 + N_2) + (N_4 + N^*_4) \big] \\
&\quad + \big[(N_{-1} + 5N^*_{-1} + N_1 + 5N_1 + 6N^*_2 + 8N^*_3 + 3N_4 + 3N^*_4) \big] \big\} \\
&\geq N(N-1) + N(N-2) = 2N^2 - 3N.
\end{aligned}
$$

$\square$

Corollary 2.4.6. *A SPN with cost $2N^2 - 3N$ can only use standard-path crossbars of types 2 and 3.*

Halpenny–Smyth gave a construction of binary SPNs with the minimum cost $2N^3 - 3N$. First consider the even N case. Construct $N/2$ pairs of paths $p_k = \{(i_{2k-1}, j_{2k-1}), (i_{2k}, j_{2k})\}$. Let every two pairs intersect twice and put eight crossbars of type 2 at the eight intersection points. Put a crossbar of type 3 at the middle of each pair p_k to connect the two paths in it. Figure 2.4.3 illustrates such a construction with $N = 6$.

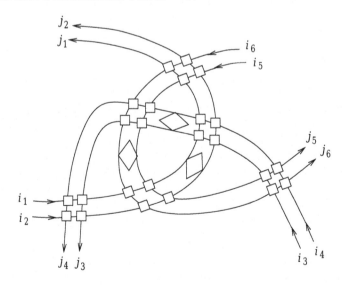

Figure 2.4.3: A SPN with $N = 6$

Consider the standard path of (i, j). If i, j are in the same p_k, then i connects to j through p_k and the type-3 crossbar on it. If i is in p_k and j in p_ℓ, then i should turn to the path of j the first time path i crosses path j. It is easily verified that two standard paths (i, j), (i', j') are link-disjoint if $i \neq i'$ and $j \neq j'$.

The number of crossbars is

$$8\binom{N/2}{2} + N/2 = N(N-2) + \frac{N}{2} = N^2 - \left\lfloor \frac{3N}{2} \right\rfloor .$$

If N is odd, simply delete one path and the type-3 crossbar on it. The number of crossbars is

$$(N+1)^2 - \frac{3(N+1)}{2} - 4\left[\frac{(N+1)^2}{2} - 1\right] - 1 = N^2 - \frac{3N-1}{2} = N^2 - \left\lfloor \frac{3N}{2} \right\rfloor .$$

Using similar but more involved arguments Halpenny–Smyth also proved that a binary staged network has at least $N^2 - 2\lfloor N/2 \rfloor - 1$ crossbars and a planar binary-staged network has at least $N^2 - \lfloor N/2 \rfloor - 2$ switches. They gave constructions of networks either matching these lower bounds or missing by one crossbar. However, as these lower bounds are all of $\Omega(N^2)$ order, they have essentially the same cost as the crossbar, which, of course, is a SNB by specifying the (i, j) path to be the unique one-turn path (turning at the point with coordinates (i, j)).

References

Arora, S., Leighton, T., & Maggs, B. 1990. On-line algorithms for path selection in a nonblocking network. *ACM Symp. Thy. Comput.*, **22**, 149–158.

Bassalygo, L. A., & Pinsker, M. C. 1974. Complexity of an optimum nonblocking switching network without reconnections. *Probl. Inform. Transm.*, **9**, 64–66.

Beneš, V. E. 1965. *Mathematical Theory of Connecting Networks and Telephone Traffic.* New York: Academic.

Cantor, D. G. 1971. On nonblocking switching networks. *Networks*, **1**, 367–377.

Chang, F. H., Guo, J. Y., Hwang, F. K., & Lin, C. K. 2004. Wide-sense nonblocking for symmetric or asymmetric 3-stage Clos networks under various routing strategies, Theor. *Comput. Sci.*, **314**, 375–386.

Clos, C. 1953. A study of non-blocking switching networks. *Bell Syst. Tech. J.*, **32**, 406–424.

Du, D. Z., Fishburn, P., Gao, B., & Hwang, F. K. 2001. Wide-sense nonblocking for 3-stage Clos networks, in Switching Networks: Recent Advances, Eds:D. Z. Du and H. Q.-Ngo, Kluwer, Amsterdam, pp. 89–100.

Feldman, P., Friedman, J. & Pippenger, N. 1986. Nonblocking networks. *ACM Symp. Thy. Comput.*, **15**, 42–54.

Friedman, J. 1988. A lower bound on strictly nonblocking networks. *Combinatorica*, **8**, 185–188.

Gabber, O., & Galil, Z. 1981. Explicit construction of linear-sized superconcentrators. *J. Comput. Syst. Sci.*, **22**, 407–420.

Halpenny, L. 1990. Nonblocking staged and repeated stage networks. Unpublished manuscript.

Halpenny, L., & Smith, C. 1992. A classification of minimal standard-path 2×2 switching networks. *Theor. Comput. Sci.*, **102**, 329–354.

Hollmann, L., & van Lint, Jr., J. H. 1992. Nonblocking self-routing switching networks. *Disc. Appl. Math.*, **37/38**, 319–340.

Hwang, F. K. 1998. Choosing the best $\text{Log}(N, m, p)$ strictly nonblocking networks. *IEEE Trans. Commun.*, **46**, 454–455.

Jimbo, S., & Maruoka, A. 1985. Expenders obtained from affine transformations. *ACM Symp. Thy. Comput.*, **17**, 88–97.

Kharkevich, A. D. 1957. Multistage construction of totally accessible communication systems. *D.A.N. SSSR*, **112**, 1043–1046.

Kurshan, R. P., & Beneš, V. E. 1980. Wide-sense nonblocking network made of square switches. *Elec. Lett.*, **17**, 697–700.

Lea, C.-T. 1991. Multi-$\log_2 N$ networks and their applications in high-speed electronic and photonic switching systems. *IEEE Trans. Commun.*, **38**, 1740–1749.

Lin, G., & Pippenger, N. 1994. Parallel algorithms for routing in nonblocking networks. *Math. Syst. Thy.*, **27**, 29–40.

Lubotzky, A., Phillips, R., & Sarnak, P. 1986. Explicit expanders and Ramanujan conjectures. *ACM Symp. Thy. Comput.*, **17**, 240–246.

Mantel, W. 1907. Solution of Problem 28. *Wiskundige Opgaven*, **10**, 60–61.

Margulis, G. A., 1793. Explicit constructions of concentrators. *Prob. Inform. Transm.*, **9**, 325–332.

Pippenger, N. 1978. On rearrangeable and non-blocking switching networks. *J. Comput. Syst. Sci.*, **17**, 145–162.

Pippenger, N. 1982. Telephone switching networks. *Amer. Math. Month.*, **26**, 101–132.

Richards, G. W., & Hwang, F. K. 1999. Extended generalized shuffle networks: sufficient conditions for strictly nonblocking operation. *Networks.*, **33**, 269–291.

Shannon, C. E. 1950. Memory requirements in a telephone exchange. *Bell Syst. Tech. J.*, **29**, 343–349.

Shyy, D.-J., & Lea, C.-T. 1991. $\text{Log}_2(N, m, p)$ strictly nonblocking networks. *IEEE Trans. Commun.*, **39**, 1502–1510.

Smith, D. G. 1977. Lower bound on the size of a 3-stage wide-sense nonblocking network. *Elec. Lett.*, **13**, 215–216.

Smyth, C. Y. 1990. Bounds for non-blocking switch networks. *Pages 260–263 of:* Luck, J. M., Moussa, P., & Waldschmidt, M. (eds), *Number Theory and Physics*, vol. 47.

Tsai, K. H., Wang, D. W., & Hwang, F. K. 2001. Lower bounds of wide-sense nonblocking Clos networks. Theor. Comput. Sci. **261**, 323–328.

Upfal, E. 1989. An $O(N \log N)$ deterministic packet routing scheme. *Proc. 21st ACM Symp. Thy. Comput.*, 241–250.

Yang, Y. Y. and Wang, W. 1999. Wide-sense nonblocking Clos networks under packing strategy, IEEE Trans. Comput. **45**, 265–284.

Chapter 3. Rearrangeable Networks

Rearrangeable networks are designed for scheduled traffic, namely, the problem is to route a frame F of requests simultaneously, which is equivalent to an edge-coloring of $G(F)$. Note that graph theory is not particularly suitable for nonblocking networks which are designed for dynamic traffic since the graph defined would be changing all the time with the additions and deletions of connections.

3.1 3-stage Clos Network

König (1916) proved the following well-known result:

Lemma 3.1.1. *Consider a bipartite graph of maximum degree d. Then d colors are necessary and sufficient for edge-coloring.*

The fundamental theorem of rearrangeability follows immediately.

Theorem 3.1.2. *$C(n_1, r_1, m, n_2, r_2)$ is rearrangeable if and only if $m \geq \max\{n_1, n_2\}$.*

Proof. For any frame F, $G(F)$ has maximum degree $d \leq \max\{n_1, n_2\}$. Hence $m = \max\{n_1, n_2\}$ suffices. On the other hand, let F be a frame in which an input or output crossbar generates $\max\{n_1, n_2\}$ requests. Then $m = \max\{n_1, n_2\}$ is necessary. □

This theorem was first given by Slepian (1952) in an unpublished manuscript. Later, Duguid (1959) gave a formal proof in a progress report. Even without a journal publication, this result, now generally known as the Slepian–Duguid theorem, was made famous by Beneš 1965 classic book. The proof of Slepian–Duguid uses Hall's (1932) system of distinct representatives, which is well known to be equivalent to edge-coloring of bipartite graphs.

As pointed out in Sec. 1.3, rearrangeable networks can also be used for dynamic traffic by rearranging existing calls. In this application, the number of connections needed to be rearranged is an important statistic to be minimized.

A convenient tool to study this number is the *Paull matrix* P as a representation of a network state. The rows of P are the input crossbars, the columns the output crossbars, and cell (i,j) contains the set of middle crossbars carrying a connection from input crossbar i to output crossbar j. Note that no middle crossbar can appear more than once in a row or a column. Furthermore, an (i,j) request needs rearrangements if and only if every middle crossbar appears either in row i or column j.

Let $\varphi(n,m,r)$ denote the minimum number of rearrangements needed for $C(n,m,r)$. Paull (1962) proved that $\varphi(n,m,r) \leq 2r-2$, while Beneš (1965) improved to $r-1$.

Theorem 3.1.3. $\varphi(n,n,r) = r-1$.

Proof. Suppose the new request to be routed is from input $x \in X$ to output $z \in Z$. Since x has at most $n-1$ busy co-inputs and z at most $n-1$ busy co-outputs, there exist a middle crossbar a available to X, and a middle crossbar b available to Z.

Without loss of generality, we may assume $X = Z = 1$, row 1 in P does not contain a and column 1 does not contain b. Furthermore, we assume that row 1 contains b and column 1 contains a for otherwise we can use the missing middle crossbar (a or b) to route the $(1,1)$ request.

Starting from the a in the first column, we search the $abab\cdots$ path which satisfies

(i) a and b alternate in the path.

(ii) Every adjacent pair ab lies in the same row, and every adjacent pair ba in the same column.

The path stops when no more a or b satisfying (i) and (ii) exists. Swap a and b in the path. The property that no row or column contains more than one a and one b is preserved. Furthermore, column 1 now contains b but not a. Hence we can use a to connect the $(1,1)$ request.

Similarly, we can search the $baba\cdots$ path starting from the b in row 1. It is easily verified that these two paths are disjoint or one row or column would contain two a's or two b's. Since there is a total of $2r-2$ a's and b's, one of the two paths has length at most $r-1$. We use that path for rearranging. Hence $\varphi(n,m,r) \leq r-1$. By noting that the worst case can be realized, Theorem 3.1.3 follows.   □

Figure 3.1.1 illustrates how an ab-path is swapped such that a can be used to connect a (I_1, O_1) request.

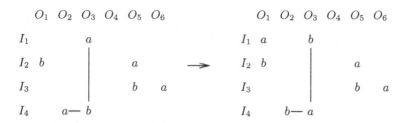

Figure 3.1.1: Swapping a path.

Bassalygo (1973) attempted to reduce the number of rearrangements provided there exists a third middle crossbar c carrying s connections with s small. Let $\varphi(n, m, r; s)$ denote the minimum number of rearrangements for $C(n, m, r)$ under this condition. Let $\varphi_1(n, m, r; s)$ denote the same except with the further restriction that all connections rearranged lie on one path of the Paull matrix. Bassalygo claimed the following result with only a sketch of proof.

Claim. $\varphi_1(n, n, r; s) \leq s + \sqrt{2s} + 1$.

Hwang and Lin (2001) gave a family of examples which disproved Bassalygo's claim for $s \geq 12$. Hwang and Lin also gave an upper bound $\varphi_1(n, n, r; s) \leq 2s$ and conjectured that $3s/2$ is the ultimate upper bound.

With more middle crossbars, i.e., $m > n$, one expects the required number of rearrangements to reduce. However, $\varphi(n, m, r)$ remains an open problem for general $m, r < m < 2n - 1$. Only the extreme case $m = 2n - 2$ was solved by Bassalygo–Grushko–Neiman (1970).

Theorem 3.1.4. $\varphi(n, 2n - 2, r) = \lfloor \left(\log_2 \frac{r(n-2)+1}{2n-3} \right) / \log_2(n - 1) \rfloor + 1$.

Proof. Let $r(j)$ denote the minimum r such that $\varphi(n, 2n - 2, r) = j$. Then

$$r(j) = \sum_{i=0}^{j-1}(n - 1)^i + (n - 1)^{j-1} \text{ for } j = 1, 2, \cdots.$$

We will prove the equation by constructing matrix $P(j)$ for $r(j)$. Since the construction is symmetric with respect to rows and columns, we will describe only in one direction. Divide the columns into subsets $C_1(j), \ldots, C_j(j)$, where $C_i(j)$ represents the set of columns necessary to force i rearrangements.

Suppose the blocked request is in cell $(1, 1)$. Then row 1 and column 1 must contain the whole set of $2n - 2$ middle crossbars. Let R and C denote the set of middle crossbars in row 1 and column 1 respectively. Since $|R| \leq n - 1$ and $|C| \leq n - 1$, necessarily, $R \cap C = \emptyset$. Hence none of $R \cup C$ can be in cell $(1,1)$. This shows $P(1)$ has dimension 2 with all R in cell $(1,2)$, and C in cell $(2,1)$.

To construct $P(2)$, then each crossbar in R must have the whole set C in the same column to force two rearrangements. But $|C| = n - 1$, hence the crossbars in row 1 must occupy distinct cells, which leads to $|C_1(2)| = n - 1$. Furthermore, since a crossbar of R will appear in every column headed by these crossbars of C, it has to take up $n - 1$ rows in $|R_2(2)|$. On the other hand we can pack the $n - 1$ appearances of every crossbar of C in an $(n - 1) \times (n - 1)$ square by using an orthogonal Latin square of order $n - 1$. Hence $|C_2(2)| = |R_2(2)| = n - 1$.

To construct $P(3)$, then the $(n - 1)^2$ total appearance of crossbars of C must each occupy a distinct row for the whole set R must appear in each such row. This leads to $|C_2(3)| = |R_2(3)| = (n - 1)^2$. But the $(n - 1)^3$ total appearances of crossbars of R can be packed into $n-1$ orthogonal Latin squares of order $n - 1$. Hence $|C_3(3)| = (n - 1)^2$.

Similar constructions work for each $P(j)$. An example of $P(3)$ with $n = 3$ is given in Fig. 3.1.2. Finally,

$$
\begin{aligned}
r_j &= \sum_{i=0}^{j-1}(n - 1)^i + (n - 1)^{j-1} = \frac{(n - 1)^j - 1}{(n - 1) - 1} + (n - 1)^{j-1} \\
&= \frac{2(n - 1)^j - (n - 1)^{j-1} - 1}{n - 2} = \frac{(n - 1)^{j-1}(2n - 3) - 1}{n - 2}.
\end{aligned}
$$

Theorem 3.1.4 follows immediately.   □

3.2   Routing Algorithms for C(n, m, r)

There are three general approaches for designing a routing algorithm:

(i) *Matrix decomposition.* Represent a frame as a matrix M where rows are input crossbars, columns are output crossbars, and the entry in cell (i, j) indicates the number of (i, j) requests. The problem is to partition M into a set of "permutation" matrices where each row and each column has at most one entry of "1". Note that all requests in a permutation matrix can be routed through one middle crossbar.

	C_1				C_2				C_3			
	a_1	a_2										
b_1			a_1	a_2								
b_2					a_1	a_2						
	b_1						a_1	a_2				
	b_2						a_2	a_1				
		b_1							a_1	a_2		
		b_2							a_2	a_1		
			b_1	b_2								
			b_2	b_1								
					b_1	b_2						
					b_2	b_1						

Figure 3.1.2: $P(3)$ with $n = 3$.

(ii) *Matching.* Treating a frame as a bipartite graph G with the input cross-bars and the output crossbars as the two parts. Recursively find a matching and remove it from G until all edges in G are gone. The edges (requests) in a matching can be routed through one middle crossbar.

(iii) *Edge coloring.* Color the edges of G so that edges incident to the same vertex have distinct colors. Edges of the same color can be routed through one middle crossbar.

Although the three settings are apparently equivalent, each may call upon intuition and techniques unique to its own. Furthermore, for problems of practical sizes, the complexity analysis of an algorithm is not the only factor to consider. First of all, the coefficient of the complexity order is usually hidden which hampers a real comparison. Secondly, one often has to introduce small variations unique to a particular application and would feel more comfortable to work with the setting the user is familiar with. For example, one who is not familiar with graph theory may prefer to write a code based on a matrix decomposition algorithm.

We will first introduce the matrix decomposition algorithm. Since there are r rows each carrying n entries, it takes $O(rn) = O(N)$ time to construct the matrix.

Waksman (1968) proposed an algorithm for $n = 2$, which was elaborated by Opferman–TsaoWu (1971) and called the *looping algorithm*. For $n = 2$, every row (every column) of M sums up to at most 2. Let a and b denote the

two middle crossbars. For those cells of M with a 2-entry, assign one request
to a and the other to b. For the remaining cells, select an arbitrary 1-entry and
assign it to a. This entry lies in a unique path (could be cyclic) of 1-entries
where two adjacent 1-entries either share a row or a column. Note that if it
is a cycle, then the length must be even. Assign a and b alternately to the
1-entries on this path. If there is a 1-entry not on this path and not assigned,
start the same process again. Since every request is examined once, the time
complexity is $O(N)$.

Andersen (1977) extended the looping algorithm to the case $n = 2^k$ for any
integer k. The idea is to apply the looping algorithm k times. By symmetry,
we will only talk about the input side. Partition the 2^t inputs of each crossbar
into 2^{t-1} sets of 2 and treat each such pair as if they were the inputs of a 2×2
crossbar. Apply the looping algorithm to assign one input of each pair to a
middle crossbar 0 and the other to a middle crossbar 1. Partition the 2^{t-1}
inputs assigned to $0(1)$ into 2^{t-2} pairs and treat each pair as if they were the
inputs of a 2×2 crossbar. Apply the looping algorithm to assign one input
of each pair to a middle crossbar $00(10)$ and another to a $01(11)$. Repeat this
procedure $t - 1$ times, then the 2^t inputs on each input crossbar is assigned
to a middle crossbar with a distinct binary number with t bits. The time
complexity is $O(Nt) = O(N \log n)$.

For general n, Bassalygo–Grushko–Neiman (1969) gave a 2-phase algorithm where backtracking is required in the second phase. Their analysis
showed the time complexity to be $Nr^3/6$. TsaoWu (1974) improved this algorithm in its first phase, which does not affect the time complexity. Ramanujan
(1973) gave a no-backtracking algorithm, but was found to be incomplete by
Kubale (1982). Jajszczyk (1985) gave a no-backtracking algorithm, but was
found to be incomplete by Cardot (1985). Gordon–Srikanthan (1990) proposed
a new matrix, called a *specification matrix*, to represent a frame by interchanging the roles of middle crossbars and output crossbars. In their matrix, rows are
input crossbars, columns are middle crossbars and cell (i, k) contains the output crossbar j if k is assigned to carry an (i, j) request. Gordon–Srikanthan
gave an algorithm which was shown to be incomplete by Chiu–Siu (1991),
whose modification was also shown to be incomplete by Lee–Hwang–Capinelli
(1997). We now present the Lee–Hwang–Capinelli algorithm which decom-
poses the specification matrix.

If a frame matrix is not of full capacity, add artificial requests such that
every row and every column sums to n. This can be done in $O(r^2)$ time by
going through each row not summing to n and using a greedy algorithm to

add artificial requests to any cell whose column is not full. Construct an initial specification matrix from the frame matrix by arbitrarily assigning to each row i a request (i, j) under column k (only j needs to be entered into cell (i, k)). This can be done in $O(N)$ time. Note that the initial specification matrix is not enforceable since the entries in a column may not be distinct, which means that a middle crossbar is assigned to carry more than one connection from the same output crossbar. The algorithm revises the specification matrix in iterations until each column has distinct entries.

A column is called e-*excessive* if it contains more than one e, and e-*deficient* if it contains no e. e is called *balanced* if no e-excessive column exists.

The Lee–Hwang–Capinelli algorithm:

Step 1. Find the smallest e such that an e-excessive column k and an e-deficient column k' exist. Suppose cell (i, k) contains e and cell (i, k') contains e'. Swap e and e' in these two cells. If $e < e'$, go back to Step 1.

Step 2. Suppose $e > e'$. Then e' is balanced before the swap. After the swap column k becomes e'-excessive and column k' e'-deficient. Let cell (i', k) contain the other e'. Swap cells (i', k) and (i', k'). Suppose cell (i', k') contains e''. If $e < e''$, go back to Step 1. If $e > e''$, then Step 2 has to be repeated. We show that e', e'', $\cdots$ are all original elements of column k', hence distinct (they are balanced elements). Suppose to the contrary that e^j is the first element in e', e'', $\cdots$ which is originally in column k, and was switched to column k' through swapping with e^i for some $i < j$. The fact that e^j is in the sequence e', e'', $\cdots$ implies $e^{j-1} = e^i$, contradicting our assumption that e^j is the first to repeat. Thus in at most e swaps, the new element e^* swapped to column k satisfies $e < e^*$, and the process goes back to Step 1.

The algorithm ends when all e are balanced. Note that whenever Step 1 is entered, either e increases or the number of e-deficient columns decreases. Hence the algorithm can enter no loop. Lee–Hwang–Capinelli gave an implementation in time $O(Nr)$. Figure 3.2.1 illustrates this algorithm for $n = 3$ and $r = 5$. The two starred entries in each matrix form the swapped pair.

For both matching algorithms and edge-coloring algorithms we need to represent a frame as a bipartite graph. This part takes $O(N)$ time which will not be counted in the following analysis.

A *maximum matching* is a matching with the maximum number of edges. Chen and Frank (1975) first proposed the use of maximum matching algorithm for routing. Hall (1967) gave an $O(Nr)$ algorithm for finding a system of

$$
\begin{array}{ccccccc}
0^*\,2^*\,4 & 2\ 0\ 4 & 2^*\,0^*\,4 & 0\ 2\ 4 & 0\ 2\ 4 & 0\ 2^*\,4^* & 0\ 4\ 2 \\
1\ 3\ 2 & 1\ 3\ 2 & 1\ 3\ 2 & 1\ 3\ 2 & 1\ 3^*\,2^* & 1\ 2\ 3 & 1\ 2\ 3 \\
0\ 3\ 3 \Rightarrow & 0^*\,3\ 3^* \Rightarrow & 3\ 3\ 0 \Rightarrow & 3\ 3\ 0 \Rightarrow & 3\ 3\ 0 \Rightarrow & 3\ 3\ 0 \Rightarrow & 3\ 3\ 0 \\
0\ 4\ 1 & 0\ 4\ 1 & 0\ 4\ 1 & 0^*\,4^*\,1 & 4\ 0\ 1 & 4\ 0\ 1 & 4\ 0\ 1 \\
2\ 1\ 4 & 2\ 1\ 4 & 2\ 1\ 4 & 2\ 1\ 4 & 2\ 1\ 4 & 2\ 1\ 4 & 2\ 1\ 4
\end{array}
$$

Figure 3.2.1: An example for the Lee–Hwang–Capinelli algorithm.

representatives, which is equivalent to a maximum matching. Hopcroft and Karp (1973) gave the fastest bipartite maximum matching algorithm which requires $O(Nr^{1/2})$ times. An M-*containing matching* is a matching covering all vertices with maximum degree. Cole and Hopcroft (1982) gave the fastest M-containing matching algorithm which requires $O(N \log r)$ time. They also showed that an even faster algorithm is possible by some preprocessing. It should be noted that the time complexities given above are for finding a single matching. A factor of n should be multiplied to find n matchings to cover all edges.

Lev–Pippenger–Valiant (1981) pointed out that any bipartite edge-coloring algorithm can be translated to a routing algorithm. Gabow (1976) proposed the following edge-coloring algorithm. Suppose that the maximum degree is n in the bipartite graph G. If n is even, partition G into two edge-disjoint subgraphs each with maximum degree $n/2$. Suppose G has N edges. Then the partition can be done in $O(N)$ time, say, by first adding edges to make all degrees even, then partitioning G into cycles and then assigning the edges on each cycle alternately to the two subgraphs. If n is odd, find one M-containing matching M and partition $G \backslash M$ into two subgraphs each with maximum degree $(n-1)/2$. Iterate this process until n matchings are obtained. Assign a distinct color to the edges of each subgraph. The fastest implementation of an edge coloring algorithm was given by Gabow–Kariv (1978) which requires $O(N \log r \log n)$ time.

By using parallel processing, more specifically, by using one processor for each connection, Lev–Pippenger–Valiant was able to reduce the time complexity by replacing N with $\log N$. We explain their algorithm for the $n = 2$ case.

Consider the bipartite graph G with r vertices in each part. Since G is 2-colorable, G consists of a set of cycles such that in each cycle the two colors must alternate. (But the coloring of different cycles are independent.) Call

the endpoints of edges *nodes*. Since each vertex has two edges, it contains
two nodes, called *siblings*. The two nodes of an edge are called *mates*. The
algorithm turns the edge-coloring problem to node-coloring. Clearly, two mates
must be in the same color, while two siblings in different colors.

The algorithm consists of iterations of two operations. One is to maintain
the sameness of colors of two mates, and the other to maintain the difference
of colors of two siblings. The general picture is: at the beginning each node is
independently assigned a cycle. By using the relations of mates and siblings,
we merge cycles and adopt the convention that the smaller index of the two
cycles will become the common index.

More specifically, at each step a node v has a label (i, j) where i represents
the index of the cycle v being currently in and j represents the color of v. At the
beginning, we simply label the $N/2$ pairs of input-nodes $(1, 1), (1, 2), (2, 1), (2, 2)$
..., $(N/2, 1), (N/2, 2)$, and the $N/2$ pairs of output-nodes $(N/2 + 1, 1), (N/2 +$
$1, 2), ..., (N, 1), (N, 2)$, where nodes with the same first label are siblings. Note
that this labeling already takes care of the difference of colors in each pair of
siblings. In step 1, the labels of each pair of mates (i, j) and (i', j') are unified. Namely, if $i < i'$, then set $i' = i$ and $j' = j$. The above relabeling may
have destroyed the difference of colors in a pair of siblings. So in step 2, the
labels of each pair of siblings $(i, j), (i', j')$ are coordinated. Namely, if $i < i'$,
then set $i' = i$ and $j' \neq j$. Now the unity of colors in a pair of mates can be
destroyed so we iterate with step 1 and 2 until no more change occurs. Since
the kth iteration identifies all vertices within distance $2^k - 1$ in a cycle, at most
$\log_2 N$ iterations are needed.

The $n = 2^t$ or general n case utilizes the $n = 2$ algorithm in the same way
as described for the single-processor case.

Nassimi and Sahni (1982) gave a similar parallel algorithm for $d = 2$ when
the processors are fully connected, and also gave algorithms for various nonfull
connection patterns of processors.

Note that if the maximum number of requests generated by each input and
output crossbar in a frame is $d < n$, then all the above algorithms can take
advantage of this fact by replacing n and N in the time complexity analyses
with d and dr. For example, the specification matrix would be of size $r \times d$
instead of $r \times n$.

Koppelman and Oruc (1994) showed that a permutation routing requires
the information of at least $(r - 3)(n/2 + 1)$ requests. Hence a sequential algorithm requires at least $O(rn) = O(N)$ time. Depending on whether parallel
routing is performed over stages, and/or over crossbars in a stage, r and n can

be reduced to $\log r$ and $\log n$.

Douglas–Oruc (1993) proved that $C(n, m, r)$ cannot be self-routing except for the case $r = 2$, whose self-routability was established earlier by Koppelman–Oruc (1989).

$C(n_1, r_1, m, n_2, r_2)$ can be routed as $C(n, m, r)$ where $n = \max\{n_1, n_2\}$ and $r = \max\{r_1, r_2\}$.

3.3   Multistage Networks

Shannon (1950) gave the information-theoretic lower bound $\Omega(N \log N)$ for the cost of a rearrangeable MIN by observing that there are $N!$ distinct permutations as candidates for frames, hence $\log N! = \Omega(N \log N)$ crossbars are needed. Bassalygo and Tsybakov (1973) gave the more specific lower bound $3N \log_3 N$ (attributed to Dobrushin). Pippenger (1980) gave the currently best lower bound (with an improvement suggested by Shearer) which is good for any network, not just MIN.

Theorem 3.3.1. *The cost of a rearrangeable network* $\geq \frac{45}{7} N \log_6 N + O(N)$.

Proof. As discussed in Sec. 1.6, a network can be represented by a digraph $G(V, I, O, E)$, where every node $v \in V \backslash (I \cup O)$ has at least two inarcs and two outarcs for otherwise the unique arc could be shrunk without affecting connectivity. Furthermore, a node v with exactly two inarcs (u, v), (u', v) and two outarcs (v, w), (v, w') can be deleted and its four arcs replaced by (u, w), (u, w'), (u', w), (u', w'). Therefore we may assume that each node in G has a total degree of at least 5.

Let $\varphi = I \to O$ be a 1–1 mapping. Let $G^*(V^*, I, E^*)$ be obtained from $G(V, I, O, E)$ by φ, and adding a loop $v \to v$ for every $v \in V \backslash (I \cup O)$. By counting such a loop as two arcs, $|E^*| \leq (7/5)|E|$.

Call a path a circuit if it starts in I, ends in O, and contains no other node in $V \cup O$. A set of circuits is called a circulant if it covers every node. Each permutation routing in G corresponds to a circulant while distinct permutations yield distinct circulants. Let C denote the set of circulants a permutation can be connected in more than one way. Then $|C| \geq N!$. Therefore it suffices to prove that

$$|E^*| \geq 9 \log_6 |C|.$$

Let M be the incidence matrix with rows and columns indexed by V^*. Then $|C| = \operatorname{Perm} M$, where Perm denotes permanent. On the other hand, let

$L_i = \sum_{j \in V^*} M(i,j)$, i.e., the row-$i$ sum. Then $|E^*| = \sum_{i \in V^*} L_i$. Minc (1963) conjectured, and Bregman (1973) proved

$$\sum_{i \in V^*} L_i \geq 9 \log_6 \text{ Perm } M, \text{ which completes the proof.}$$

$\square$

For the s-stage network, Pippenger–Yao (1982) proved the following lower bound. Actually, they proved it even if the permutation is restricted to $a_i \rightarrow b_{j+i}(\mod N)$, $i = 1, \ldots, N$, for some arbitrary j (called an N-shifter).

Theorem 3.3.2. *An N-shifter of depth s has at least $sN^{1+\frac{1}{s}}$ edges.*

Proof. Let $T_s(N)$ denote a rooted tree with N leaves and depth s. Let $P_1, \ldots, P_N$ denote its N paths, and let $d(v)$ denote the degree of vertex v. We prove by induction that

$$\Delta(T_s(N)) \equiv \sum_{j=1}^{N} \sum_{v \in P_j} d(v) \geq sN^{1+\frac{1}{s}}. \tag{$*$}$$

True for $s = 1$. For general s, suppose that the root has degree d. Let $T_s^i(N_i)$ be the rooted subtree at branch i. Then

$$\Delta(T_s(N)) = dN + \sum_{i=1}^{d} \Delta(T_s^i(N_i)) \geq dN + (s-1) \sum_{i=1}^{d} N_i^{1+\frac{1}{s-1}}$$

$$\geq dN + (s-1)d \left(\frac{N}{d}\right)^{1+\frac{1}{s-1}} = dN + (s-1)N^{1+\frac{1}{s-1}} d^{-\frac{1}{s-1}}.$$

Minimizing over d, set

$$\frac{\partial \Delta}{\partial d} = N + (s-1)N^{1+\frac{1}{s-1}} \left(-\frac{1}{s-1}\right) d^{-\frac{1}{s-1}-1} = N - \left(\frac{N}{d}\right)^{1+\frac{1}{s-1}} = 0.$$

Thus $d^\circ = N^{\frac{1}{s}}$. Replacing d with d° in the above $\Delta(T_s(N))$, we obtain $(*)$.

Let P_{ij} denote the path from input i to output $j + i$ in an N-shifter $G(V, I, O, E)$. Assemble P_{ij}, $j = 1, \ldots, N$, into an N-tree T_i by identifying common initial segments of two paths (but once they split, they remain split even if they have common segments later). By $(*)$,

$$\Delta(T_i) \geq sN^{1+\frac{1}{s}}.$$

Let $\mu(i, j, e)$ be 1 if the arc e is from a node in P_{ij} in $G(V, I, O, E)$. Then

$$\sum_{e \in E} \mu(i, j, e) \geq \sum_{v \in P_{ij}} d_{T_i}(v).$$

(The strict inequality holds if a node v appears in both paths P_{ij} and $P_{ij'}$ after the split, then (v, w) on $P_{ij'}$ does not contribute to $d_{T_i}(v)$ for v in P_{ij}.) Summing over j, we have

$$\sum_{1 \leq j \leq N} \sum_{e \in E} \mu(i, j, e) \geq \sum_{1 \leq j \leq N} \sum_{v \in P_{ij}} d_{T_i}(v) = \Delta(T_i) \geq sN^{1+\frac{1}{s}},$$

or

$$\sum_{1 \leq i \leq N} \sum_{1 \leq j \leq N} \sum_{e \in E} \mu(i, j, e) \geq sN^{2+\frac{1}{s}}.$$

On the other hand, since the paths $P_{1j}, \ldots, P_{Nj}$ form a permutation routing and hence have no common nodes,

$$\sum_{1 \leq i \leq N} \mu(i, j, e) \leq 1,$$

or

$$\sum_{1 \leq i \leq N} \sum_{1 \leq j \leq N} \sum_{e \in E} \mu(i, j, e) \leq N|E|.$$

Thus we have

$$|E| \geq sN^{1+\frac{1}{s}}.$$

$\square$

Pippenger–Yao also gave a probabilistic argument that an s-stage rearrangeable network with cost $O\left(sN^{1+\frac{1}{s}}(\log N)^{\frac{1}{s}}\right)$ exists.

Spanke–Beneš (1987) gave (they credited the argument to R. L. Graham)

Theorem 3.3.3. *The number of 2×2 crossbars in a planar rearrangeable binary staged-network is at least $\binom{N}{2}$.*

Proof. Consider the frame where input i is paired with output $N + 1 - i$. Then any two paths must cross at a different crossbar. Hence $\binom{N}{2}$ crossbars are needed. $\square$

Kautz–Levitt–Waksman (1968) gave a construction achieving the above lower bound.

Next we give some deterministic constructions of rearrangeable MINs.

In Theorem 3.1.2, the only property of the crossbar that was used to prove the rearrangeability is that a crossbar itself is rearrangeable. Namely, any matching between the inlets and the outlets of a crossbar can be simultaneously connected. Therefore, if we replace a crossbar with a rearrangeable network of the same size, then the rearrangeability of the network is preserved.

Consider the sequence of networks ν_k, $k = 1, \ldots, K$, obtained by the above recursive construction where $\nu_1 = [X_{a_1 b_1}, X_{a_0, a_0}, X_{b_1 a_1}]$ and

$$\nu_k = [X_{a_k b_k}, \nu_{k-1}, X_{b_k a_k}] \quad \text{for} \quad k = 2, \ldots, K.$$

Pippenger (1978) called ν_k a *sandwich network* and proved

Theorem 3.3.4. *The minimum cost of a rearrangeable* $\nu_K = 2(K+1)\left(\frac{N^{K+2}}{2}\right)^{\frac{1}{K+1}}$, *where* $N = \prod_{i=0}^{K} a_i$.

Proof. We first prove that

$$\text{Cost}(\nu_K) = 2 \sum_{1 \leq k \leq K} \left(\prod_{1 \leq i \leq k} a_i \right) \left(\prod_{k \leq j \leq K} b_j \right) a_0 + \left(\prod_{1 \leq j \leq K} b_j \right) a_0^2 \qquad (*)$$

by induction on K. $(*)$ is easily checked for $K = 1$. For general K,

$$\text{Cost}(\nu_K) = 2 a_K b_K \left(\prod_{1 \leq k \leq K-1} a_k \right)$$

$$+ b_K \left[2 \sum_{1 \leq k \leq K-1} \left(\prod_{1 \leq i \leq k} a_i \right) \left(\prod_{k \leq j \leq K-1} b_j \right) a_0 + \left(\prod_{1 \leq j \leq K-1} b_j \right) a_0^2 \right]$$

$$= 2 b_K \left(\prod_{1 \leq k \leq K} a_k \right) + 2 \sum_{1 \leq k \leq K-1} \left(\prod_{1 \leq i \leq k} a_i \right) \left(\prod_{k \leq j \leq K} b_j \right) a_0 + \left(\prod_{1 \leq j \leq K} b_j \right) a_0^2$$

$$= 2 \sum_{1 \leq k \leq K} \left(\prod_{1 \leq i \leq k} a_i \right) \left(\prod_{k \leq j \leq K} b_j \right) a_0 + \left(\prod_{1 \leq j \leq k} b_j \right) a_0^2.$$

By Theorem 3.1.2, ν_K is rearrangeable if $a_i \leq b_i$ for all $1 \leq i \leq K$. Set $a_i = b_i$. Then

$$\text{Cost}(\nu_K) = 2 \sum_{1 \leq k \leq K} a_k N + N a_0 = \left(2 \sum_{1 \leq k \leq K} a_k + a_0 \right) N.$$

Since $\sum_{0 \le k \le K} \log a_k = \log N$ is a constant, we use the Lagrange multiplier to minimize Cost (ν_k). Set

$$f = \left(2 \sum_{1 \le k \le K} a_k + a_0 \right) N - \lambda \sum_{k=0}^{K} \log a_k \,.$$

Setting

$$\frac{\partial f}{\partial a_k} = 2N - \frac{\lambda}{a_k} = 0 \quad \text{for} \quad k \ne 0$$

and

$$\frac{\partial f}{\partial a_0} = N - \frac{\lambda}{a_0} = 0 \,,$$

we obtain $a_1 = a_2 = \cdots = a_K = a_0/2$. Hence $a_0 (a_0/2)^K = N$, or $a_0 = 2(N/2)^{\frac{1}{K+1}}$ and $a_k = (N/2)^{\frac{1}{K+1}}$ for $1 \le k \le K$. Theorem 3.3.4 follows immediately. $\qquad\square$

Corollary 3.3.5. $Cost(\nu_K) \to 6N \log_3 N + O(N)$ as $K \to \infty$.

Proof. Since

$$\frac{a}{3 \log_3 a} \ge \frac{3}{3 \log_3 3} = 1 \,,$$

$$\text{Cost}(\nu_k) \ge \left(2 \sum_{1 \le k \le K} 3 \log_3 a_k + a_0 \right) N$$

$$= 6(\log_3 N - \log_3 a_0 + a_0)N = 6N \log_3 N + O(N) \,.$$

$\qquad\square$

ν_k is called a *d-nary Beneš network* if $a_k = d$ for $0 \le k \le K$ (the binary Beneš network is also known simply as *Beneš network*). For a $(2K+1)$-stage d-nary Beneš network, $N = d^{K+1}$, or $K = \log_d N - 1$

$$\text{Cost}(\nu_K) = \left(2 \sum_{1 \le k \le K} a_k + a_0 \right) N = [2d(\log_d N - 1) + d\,]N = 2dN \log_d N + O(N).$$

Hence the tertiary Beneš network achieves the upper bound $6N \log_3 N$ for sandwich networks.

Let ν be a $\log_2 N$-stage binary network. Again we ask how many stages of horizontal extension, or how many copies of vertical stacking of ν is sufficient for rearrangeability. Since a 2×2 crossbar is a binary device and there are N distinct permutations, we need at least

$$\log_2(N!) \sim N \log_2 N - 1.44N + \frac{1}{2} \log_2 N$$

2×2 crossbars. On the other hand, a $\log_2 N$-stage network has $N \log_2 N/2$ crossbars, while each stage has $N/2$ of them. A simple counting argument shows that the horizontal extension needs at least $2 \log_2 N - 2$ stages, while the vertical stacking needs at least 2 copies. It turns out that the first bound is much closer to the truth than the second.

As we discussed in Sec. 1.5, there are many ways to add extra stages to a buddy network. And one does not expect a unified answer to the above question from all these varieties of networks and additions. Thus it is sensible to study a specific choice. It is very natural to choose the Omega network since its linking pattern is invariant from stage to stage which greatly reduces the ambiguity of how to add the extra stages. It is of particular interest to keep the same shuffle-exchange linking pattern for all stages (so that the components of the network are modular) and ask how many stages suffice. We first give a lower bound which is better than the information-theoretic lower bound $\lceil \frac{\log_2(N!)}{N_2} \rceil = 2 \log_2 N - 2$.

For $N = 2^n$, Linial and Tarsi (1989) defined an $N \times (n + s)$ binary matrix to be *balanced* if every n adjacent columns yield a submatrix such that the binary numbers in the N rows are all distinct. Call the submatrix B_j if it starts with column j. Figure 3.3.1 illustrates an 8×6 balanced matrix.

$$
\begin{array}{cccccc}
0 & 0 & 0 & 1 & 1 & 1 \\
0 & 0 & 1 & 0 & 1 & 1 \\
0 & 1 & 0 & 0 & 0 & 1 \\
0 & 1 & 1 & 0 & 0 & 0 \\
1 & 0 & 0 & 0 & 1 & 0 \\
1 & 0 & 1 & 1 & 0 & 0 \\
1 & 1 & 0 & 1 & 0 & 1 \\
1 & 1 & 1 & 1 & 1 & 0 \\
\end{array}
$$

Figure 3.3.1: An 8×6 balanced matrix

Lemma 3.3.6. *An N-permutation π can be routed through an s-stage shuffle exchange network if and only if there exists an $N \times (n + s)$ balanced matrix such that if the binary number in row i of B_1 is b_i, then the binary number in row i of B_{s-n+1} is $\pi(b_i)$.*

Proof. In the shuffle-exchange linking pattern, the inlink labeled $(x_1, x_2, \ldots, x_n)$ either connects to $(x_2, \ldots, x_n, 0)$ or $(x_2, \ldots, x_n, 1)$. Furthermore, the two inlinks (of the same crossbar) having the same $x_1, \ldots, x_{n-1}$, must one go to $(x_2, \ldots, x_n, 0)$ and the other to $(x_2, \ldots, x_n, 1)$ to avoid conflict. By interpreting row i in B_j as the label of the stage-$(j-1)$ output link on the path $(b_i, \pi(b_i))$, and recalling the shuffle-exchange linking pattern as remarked above, there exists a 1–1 mapping between successful routings of π and a balanced matrix with $\{b_i\}$ and $\pi\{b_i\}$ in the same row. $\qquad\square$

Varma and Raghavendra (1988) commented that no $N \times (2n - 2)$ balanced matrix exists for the bit-reverse permutation, i.e., $\pi : (x_1, \ldots, x_n) \rightarrow (x_n, \ldots, x_1)$ since B_n, having its first column identical to its last, cannot be balanced. Therefore the shuffle-exchange network needs at least $2n - 1$ stages for rearrangeability.

That $2n - 1$ shuffle-exchange stages also suffice was first conjectured by Beneš (1965). There has been a slow progress towards proving the conjecture. Stone (1971) proved that n^2 stages suffice. Lang (1976) proved that $O(\sqrt{N})$ stages suffices. Parker (1980) first obtained a linear estimate $3n$, which was improved to $3n - 1$ by Wu–Feng (1981), to $3n - 3$ by Huang–Tripathi (1986), and finally to $3n - 4$ by Varma–Raghavendra (1988). On the other hand, the conjecture has been verified for small n, by Parker (1980) for $n = 3$ and by Raghavendra (1994) for $n = 4$, with computer-aided computations. Raghavendra–Varma (1987) also gave an analytic proof for $n = 3$.

Lea (1991) considered the vertical stacking of buddy networks.

Theorem 3.3.7. *$Log_2(N, 0, m)$ is rearrangeable if and only if $m \geq 2^{\lfloor n/2 \rfloor}$.*

Proof. Since a buddy network is a unique-path network, a frame F determines a routing R_F (may contain conflicts). Define the shell-i intersection graph G_i by taking all paths in R_F as vertices, and pairs of paths intersecting each other at shell-i links as edges. Then G_i has maximum degree 2 since a 2×2 crossbar allows at most one path to intersect at a given stage. Mark an edge "front" or "back" depending on the intersection occurring at a stage-i link or at a stage-$(n - i)$ link. Since every vertex with two edges must have one marked by "front" and the other by "back", G_i contains no odd cycle and is a bipartite graph. Hence G_i can be 2-colored.

Partition the paths into two sets according to the colors of G_1. Then partition each subset into two sets according to the colors of G_2, and so on. After $\lfloor n/2 \rfloor$ partitions, the paths are distributed to $2^{\lfloor n/2 \rfloor}$ sets where each set can be routed by one copy of a buddy network since there is no intersection at any stage. Therefore $2^{\lfloor n/2 \rfloor}$ copies suffice. On the other hand, there exist a set of $2^{\lfloor n/2 \rfloor}$ inputs and $2^{\lfloor n/2 \rfloor}$ outputs such that all paths between them go through a unique node at stage $\lfloor n/2 \rfloor$. Therefore that many copies are needed. $\qquad\qquad\qquad\qquad\qquad\qquad\qquad\qquad\qquad\qquad\qquad\qquad\quad$ □

Lea–Shyy (1991) extended the sufficiency part of Theorem 3.3.7 to $\log_2(N, k, m)$ networks.

Note that Theorem 3.3.7 cannot be straightforwardly extended to the d-nary version since the maximum degree in G_i is $2d - 2 > d$ for $d > 2$. Hwang and Lin (2003) gave a different proof for the d-nary version.

Theorem 3.3.8. *$Log_d(d^n, 0, m)$ is rearrangeable if and only if $m \geq d^{\lfloor n/2 \rfloor}$.*

Proof. Sufficiency.   Thoerem 3.3.8 is obviously true for $n = 1$ (the network is just a $d \times d$ crossbar) and $n = 2$ (at most d requests can use the same link). We prove the general $n \geq 3$ case by induction.

Take a $BY_d^{-1}(n, 0)$ and remove its first and last stages. Then the network is reduced to d^2 copies of $BY_d^{-1}(n - 2, 0)$. Since $BY_d^{-1}(n, 0)$ has a unique path for any request, we can replace the source (destination) of a request from a stage-1 input (stage-n output) to a stage-2 input (stage-$(n - 1)$ output). Further, since each stage-1 (stage-n) crossbar is involved in at most d requests, each stage-2 input (stage-$(n - 1)$ output) is involved in at most d requests. For each of the $d^2 BY_d^{-1}(n - 2, 0)$, construct a bipartite graph by treating its inputs and outputs as the two parts, and add an edge (x, y) if there is a request from x to y. This bipartite graph is d-edge-colorable since its degrees are bounded by d. Thus we can spread the requests in each $BY_d^{-1}(n - 2, 0)$ into d copies so that in each copy the requests are regular point-to-point, i.e., each input and output is involved in at most one request. By induction, each such copy is rearrangeable if the requests can be further spread to $d^{\lfloor (n-2)/2 \rfloor}$ copies. Therefore a total of $d \times d^{\lfloor (n-2)/2 \rfloor} = d^{\lfloor n/2 \rfloor}$ copies suffice.

Necessity.   Fix a link x at link-stage $\lfloor n/2 \rfloor$ in $BY_d^{-1}(n, 0)$. Let $X(Y)$ denote the set of inputs (outputs) of $BY_d^{-1}(n, 0)$ which has a path to x. It is easy to verify that $|X| = d^{\lfloor n/2 \rfloor}$ and $|Y| = d^{\lceil n/2 \rceil}$. Since $BY_d^{-1}(n, 0)$ is a unique-path network, a path starting from X to Y must go through x. By assigning $d^{\lfloor n/2 \rfloor}$ requests from X to Y, we conclude that these requests are routable only if they can be spread to at least $d^{\lfloor n/2 \rfloor}$ copies of $BY_d^{-1}(n, 0)$. □

Theorem 3.3.9. *$Log_d(d^n, k, m)$ is RNB if and only if $m \geq d^{\lfloor (n-k)/2 \rfloor}$.*

Proof. Sufficiency.   Theorem 3.3.9 holds for $k = 0$ (Theorem 3.3.8). We prove the general $k \geq 1$ case by induction.

Take a $BY_d^{-1}(n,k)$ and remove its first and last stages. Then the network is reduced to d copies of $BY_d^{-1}(n-1,k-1)$ where each stage-1 and stage-n crossbar has a link to each $BY_d^{-1}(n-1,k-1)$. Note that we cannot reassign the sources and destinations to the reduced network as we did in Theorem 3.3.8 since the path of a request is not unique.

Consider a $BY_d^{-1}(n,k)$ and construct a bipartite graph G by taking the stage-1 crossbar and the stage-n crossbars as the two parts of vertices, and a request from vertex u to vertex v as an edge. Then the maximum degree of G is bounded by d and G can be d-edge-colored. Associate a color to each of the d copies of $BY_d^{-1}(n-1,k-1)$ mentioned in the above paragraph, and route the requests through these $BY_d^{-1}(n-1,k-1)$ according to their colors. This is possible since each crossbar in stage-1 and stage-n has a link to each $BY_d^{-1}(n-1,k-1)$. Then $Log_d(d^n,k,m)$ is RNB since $Log_d(d^{n-1},k-1,m)$ is RNB by induction.

Necessity.   By removing the first k stages and the last k stages of $BY_d^{-1}(n,k)$, the network is reduced to d^k copies of $BY_d^{-1}(n-k,0)$ whose inputs and outputs will be referred to as subinputs and suboutputs. For $BY_d^{-1}(n-k,0)$, fix a link-stage-$\lfloor(n-k)/2\rfloor$ link x. Let x_i be the link corresponding to x in the ith $BY_d^{-1}(n-k,0)$. Let X_i and Y_i assume the role of X and Y in the necessary part of Theorem 3.3.8. Then $|X_i| = d^{\lfloor(n-k)/2\rfloor} \equiv q$ and $|Y_i| = d^{\lceil(n-k)/2\rceil}$. Let $X_i = (x_{i1},\cdots,x_{iq})$ for $1 \le i \le d^k$. It is easily verified that x_{ij} has access to the same set I_j of d^k inputs for all i. Since each input has access to d^k subinputs, and x_{ij} already has access to the d^k subinputs in I_j, it cannot have access to any other subinput. This holds for all $1 \le j \le q$. When all inputs in $\bigcup_{j=1}^q I_j$ generate requests, then there is a total of $d^{\lfloor(n-k)/2\rfloor}d^k = d^{\lfloor(n+k)/2\rfloor}$ requests reaching X (and going to Y). By Theorem 3.3.8 each such request go through link x_i for some i so at least one x_i has to be traversed $d^{\lfloor(n+k)/2\rfloor}/d^k = d^{\lfloor(n-k)/2\rfloor}$ times, i.e., $d^{(n-k)/2}$ copies of $BY^{-1}(n,k)$ are required. $\square$

3.4   Routing of MIN

A $(2\log N - 1)$-stage Beneš network is recursively constructed from replacing a stage of crossbars with 3-stage Clos networks ($n-1$ times). Its routing algorithm is also obtained by doing the 3-stage routing algorithm $n-1$ times. Therefore, its time complexity is simply the time complexity of the 3-stage Clos network multiplied by a factor of $n = \log N$. For binary Beneš network, the complexity is $O(N\log N)$. Lee (1985) gave a routing algorithm which allows self-routing in the second half of the Beneš network. But the over-all time complexity is not affected.

For the d-nary version, if d is a power of 2 and treated as fixed, then recursively using the 3-stage Clos algorithm $\log N$ times leads to an $O(N \log N)$ algorithm. Note that the same time complexity holds if d varies from stage to stage in a symmetric network but remains a power of 2. For arbitrary $d_1, \ldots, d_k$ with $\prod_{i=1}^{k} d_i = N$, the time complexity is $O(N(\log N)^2)$.

Although the $\log_2(N, 0, 2^{\lfloor n/2 \rfloor})$ network uses the self-routing buddy network as the basic component, the routing requires solving the 2-coloring problem $\lfloor n/2 \rfloor$ times and hence requires $O(Nn) = O(N \log N)$ time. The log_2 $(N, m, 2^{\lfloor \frac{n-m}{2} \rfloor})$ network uses the extra-stage buddy network, which is not self-routing, as the basic component. Its routing requires solving the 2-coloring problem $(n + m)/2$ times, hence requires $O(N(n + m))$ time.

Since the above algorithms all require at least $O(N \log N)$ time, we have to resort to two other approaches for faster algorithms. The first is to use parallel processing, and the second is to use self-routing, but blocking, networks.

By using either the Lev–Pippenger–Valiant parallel algorithm or the Nassimi–Sahni's parallel algorithm on 3-stage Clos network n times, we obtain an $O((\log N)^2)$ algorithm for the rearrangeable Beneš network. For d-nary Beneš network, d fixed but a power of 2, extension of Lev–Pippenger–Valiant parallel algorithm on 3-stage Clos network yields an $O((\log N)^2)$ time algorithm. This could be extended to the case that $d_1, \ldots, d_k$ can be different but each a power of 2. For arbitrary $d_1, \ldots, d_k$, the time complexity is $O((\log N)^3)$.

A self-routing algorithm for a $O(\log N)$-stage network takes only $O(\log N)$ time, thus is faster than a parallel algorithm. Lenfant (1978) first showed that the Beneš network can self-route all five families of frequently used permutations, each with a different routing algorithm.

Nassimi–Sahni (1980) presented a unified self-routing algorithm for the Beneš network as the following. At stage i, $1 \leq i \leq n - 1$, the $(n + 1 - i)$-st bit of the destination of the upper input at a crossbar determines the setting of that crossbar, i.e., if the bit is 0(1), the upper input is connected to the upper (lower) output. For stage i, $n \leq i \leq 2n - 1$, use the standard self-routing, i.e., a path is controlled by bit i of its destination. Nassimi–Sahni showed that all five families of Lenfant are routable by their algorithm. Figure 3.4.1 illustrates how the Nassimi–Sahni algorithm routes the bit-reversal permutation.

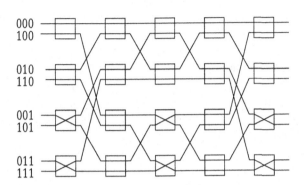

Figure 3.4.1: The Nassimi–Sahni routing.

Raghavendra–Boppana (1991) gave a self-routing algorithm which differs from the Nassimi–Sahni algorithm in letting a bit (the same bit as in the Nassimi–Sahni algorithm) of the smaller destination of the two inputs deciding the setting of a crossbar for the first $n-1$ stages, and use the standard self-routing for the last n stages.

Call a permutation π *linear* if there exists an $N \times N$ binary matrix Q such that $\pi(x) = Qx$ for every input x, and *linear-complement* if $\pi(x) = Qx + c$ for some constant vector c. Peace (1977) showed that a linear permutation π can be routed by $\Omega \circ \Omega$ by decomposing π into two simpler permutations, each routed by a half of the network. Etzion–Lempel (1986) gave an $O((\log N)^2)$ time method to route a linear permutation through a $(2n-1)$-stage shuffle-exchange network. Both algorithms are not self-routing. Raghavendra–Boppana proved. Note that a linear permutation has the property that $\pi(x) - \pi(x+a)$ is independent of x. In particular, the difference of $\pi(x)$ and $\pi(x')$ for x, x' going into the same crossbar is constant over a given stage.

Theorem 3.4.1. *A linear-complement permutation can be routed by the Raghavendra–Boppana self-routing algorithm.*

Proof. Assume the Beneš network is in the form of $BL \circ BL^{-1}$. Theorem 3.4.1 is proved by induction on n. It is trivially true for $n = 1$. We prove it for general n.

Let π be a permutation from input to output. Since the inputs are labeled in the natural order of links, Q can be viewed as a permutation from link labels to destinations. As a path is identified by its destination, and at each stage we want to find out which link a path moves to, it is more convenient to consider the inverse mapping Q^{-1} from destinations to link labels. Of course, Q^{-1} is linear-complement if and only if Q is.

Let B and A denote the permutation matrices before and after a stage whose setting is controlled by the bit y_i. More specifically, consider $x = By$ and $x' = Ay$ at a crossbar X where y is the destination and $x(x')$ the link label. Then x' differs from x at most in bit i. We show that if x_i is a linear combination of elements of y, then x'_i remains so.

Let z be the other destination at X. If $y_i \neq z_i$, then clearly, $x_i = y_i$. Suppose $y_i = z_i$. Let j be the most significant bit such that $y_j \neq z_j$. Then $x_i = y_i$ if $y_j = 0$ and $x_i = y_i \oplus 1$ if $y_j = 1$; or equivalently, $x_i = y_i \oplus y_j$. Therefore, B remains to be a linear complement permutation.

Next note that any two destinations y and y' differing only in the control bit i will have one goes to the upper outlet and the other the lower. Suppose to the contrary that both go the same direction. Then $y_i \oplus y_j = y'_i \oplus y'_j = y'_i + y_j$, contradicting the assumption that $y_i \neq y'_i$.

Therefore the set of requests sent to each half of the next stage is a linear complement permutation. Induction applies. $\qquad\square$

Figure 3.4.2 illustrates how the Raghavendra–Boppana algorithm routes the linear permutation $y = Lx$, where

$$L = \begin{pmatrix} 0 & 1 & 1 \\ 0 & 0 & 1 \\ 1 & 0 & 0 \end{pmatrix}.$$

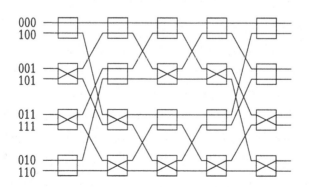

Figure 3.4.2: The Raghavendra–Boppana routing

Lee–Lu (1994) proposed a self-routing network which reduces the amount of hardware by allowing multipass of data. Their network has a cost of $O(N \log N)$ and takes $O((\log N)^3)$ time for self-routing.

A sorting network sorts a set of numbers into numerical order by using a MIN where each crossbar is a 2×2 comparator which sends the smaller of the two inlets to the upper outlet, and the larger one to the lower outlet. A sorting network of size N can be used as a self-routing rearrangeable MIN, if the frame contains N requests, by setting the destination of i as the number at input i. Since output j is the jth smallest output, the sorting will put it at the right place. Note that if some outputs are missing from the frame, then output j could be the ith smallest output for some $i < j$, and the path destined for output j will land at output i. We can fix this by adding fictitious requests to fill up the frame. Another way is to add a *shuffled* $BY^{-1}(n)$, which is an infra connector by Theorem 1.6.1, after the sorting network.

We now introduce Batcher's (1968) *bitonic sorting* network. A *bitonic sequence* is a sequence of numbers which first increase and then decrease. A *circular bitonic sequence* is a cyclic sequence which can be broken down at some point into a bitonic sequence.

Lemma 3.4.2. *Let* $(a_1, \ldots, a_{2n})$ *denote a circular bitonic sequence B. Then* $(\max\{a_1, a_{n+1}\}, \ldots, \max\{a_n, a_{2n}\})$ *consists of the n largest a_i and is circular bitonic. Similarly,* $(\min\{a_1, a_{n+1}\}, \ldots, \min\{a_n, a_{2n}\})$ *consists of the n smallest a_i and is circular bitonic.*

Proof. Without loss of generality, assume $a_k = \max\limits_{1 \le i \le 2n} a_i$. Then by the circular bitonic property, the next largest element is either a_{k-1} or a_{k+1}. In general, suppose $(a_i, a_{i+1}, \ldots, a_j)$ is the set of $j-i+1$ largest elements, then the $j-i+2$ largest element is either a_{i-1} or a_j. Hence the set L of n largest elements is consecutive. Consequently, exactly one element in each pair (a_i, a_{i+n}) belongs to L. Since L is a subsequence of a circular bitonic sequence, L is also circular bitonic. □

Lemma 3.4.3. *A shuffled $BY^{-1}(n)$ can sort a circular bitonic sequence.*

Proof. By Lemma 3.4.2, the first stage outputs two circular bitonic sequences, one consisting of the $N/2$ smallest destinations and going to the upper half of stage-2, the other $N/2$ largest destinations and going to the lower half of stage-2. Induction now applies. □

Batcher's bitonic network of size N consists of two layers of network. The ith inner network consists of 2^{n-i} copies of $BY^{-1}(i)$ lined up vertically. The outer network connects the inner networks. In partcular, a shuffle-exchange linking pattern connects two copies of $BY^{-1}(i)$ with a copy of $BY^{-1}(i+1)$. Figure 3.4.3 illustrates a bitonic network of size 8 (broken lines show stages, and heavy lines show links of the Omega network).

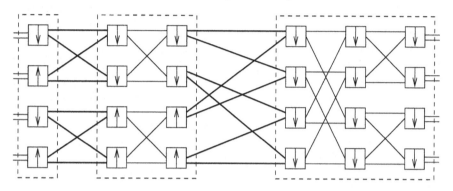

Figure 3.4.3: Batcher's bitonic network of size 8.

Theorem 3.4.4. *Batcher's bitonic network is a sorting network.*

Proof. Place all the even-numbered $BY^{-1}(i)$ for all $i = 1, \ldots, n$, upside down (see arrows in Fig. 3.4.3) so that the outputs are in reverse order. Then for $1 \leq i \leq n-1$ the inputs to $BY^{-1}(i+1)$, which are from the combined outputs of two consecutive $BY^{-1}(i)$, form a circular bitonic sequence. By Lemma 3.4.3, $BY^{-1}(i+1)$ can sort it.   □

Ajtai–Komlos–Szemerédi (1983) gave an $O(N \log N)$ sorting network which uses expander constructions and has large coefficients.

References

Ajtia, M., Komlós, J., & Szemerédi, E. 1983. An $O(n \log n)$ sorting network. *Pages 1–9 of: Proc. 15th Ann. ACM Symp. Thy. Comput.*

Andersen, S. 1977. The looping algorithm extended to base 2^t rearrangeable switching networks. *IEEE Trans. Commun.*, **25**, 1057–1063.

Bassalygo, L. A. 1973. On a number of reswitching in a three-stage connecting networks. *ITC*, **7**, 231/1–231/4.

Bassalygo, L. A., & Tsybakov, B. S. 1973. Blocking probability for a rearrangeable switching system. *Problems Inform. Transmission*, **6**, 336–348.

Bassalygo, L. A., Grushko, I. I., & Neiman, V. I. 1969. Control of two-sided rearrangeable connection networks (in Russian). *In: Studies of Probabilistic Problems in Connection Networks with Complex Structures.* Moskow: Nauka.

Bassalygo, L. A., Grushko, I. I., & Neiman, V. I. 1970. For a 3-stage connecting network. *ITC*, **6**, 241/1–241/5.

Batcher, K. E. 1968. Sorting networks and their aplications. *Proc. 1968 Spring Joint Comput. Conf.*, **32**, 307–314.

Beneš, V. E. 1965. *Mathematical Theory of Connecting Networks and Telephone Traffic.* New York: Academic.

Bregman, L. M. 1973. Certain properties of nonnegative matrices and their permanents. *Sov. Math. Dokl.*, **14**, 945–949.

Cardot, C. 1985. Comments on a simple algorithm for the control of rearrangeable switching networks. *IEEE Trans. Commun.*, **34**, 395.

Chen, C. J., & Frank, A. A. 1975. On programmable parallel data routing networks via cross-bar switches for multiple element computer architectures. *In: Goos, G., & Hartmanis, J. (eds), Parallel Processing.* New York: Springer-Verlag.

Chiu, Y. K., & Siu, W. C. 1991. Comment: Novel algorithm for Clos-type networks. *Elec. Lett.*, **27**, 524–526.

Cole, R., & Hopcroft, J. 1982. On edge coloring bipartite graphs. *SIAM J. Comput.*, **11**, 540–546.

Douglas, B. G., & Oruc, A. Y. 1993. On self-routing in Clos connection networks. *IEEE Trans. Commun.*, **41**, 121–124.

Duguid, A. M. 1959. *Structural properties of switching networks.* Progr. Rep. BTL–7. Brown Univ.

Etzion, T., & Lempel, A. 1986. An efficient algorithm for generating linear transformations in a shuffle exchange network. *SIAM J. Comput.*, **15**, 216–221.

Gabow, H. 1976. Using Euler partitions to edge coloring bipartite multigraphs. *Int. J. Comput. Inform. Sci.*, **5**, 345–355.

Gabow, H., & Kariv, O. 1978. Algorithms for edge coloring bipartite graphs. *Pages 184–192 of: Proc. 10th Ann. ACM Symp. Thy. Comput.*

Gordon, J., & Srikanthan, S. 1990. Novel algorithm for Clos-type networks. *Elec. Lett.*, **26**, 1772–1774.

Hall, M. 1967. *Combinatorial Theory.* Waltham, MA: Blaisdell.

Hall, P. 1932. Distinct representatives of subsets. *J. London Math. Soc.*, **10**, 26–30.

Hopcroft, J., & Karp, R. 1973. An $n^{5/2}$ algorithm for maximum matching in bipartite graphs. *SIAM J. Comput.*, **2**, 225–231.

Huang, A., & Knauer, S. 1984. Starlite: A wideband digital switch. *Proc. Globecom.*

Huang, S. T., & Tripathi, S. K. 1986. Finite state model and compatibility theory: New analysis tools for permutation networks. *IEEE Trans. Comput.*, **35**, 591–601.

Hwang, F. K., & Lin, W. D. 2001. The number of rearrangements in a 3-stage Clos network using an auxiliary switch. in Switching Networks: Recent Advances, eds: D. Z. Du and H. Q. Ngo, Kluwer, 179–190.

Hwang, F. K., & Lin, W. D. 2003. Necessary and sufficient conditions for rearrangeable $Log_d(N, m, p)$. Preprint.

Jajszczyk, A. 1985. A simple algorithm for the control of rearrangeable switching networks. *IEEE Trans. Commun.*, **33**, 169–171.

Kautz, W. H., Levitt, K. N., & Waksman, A. 1968. Cellular interconnection arrays. *IEEE Trans. Comput.*, **17**, 443–451.

König, D. 1916. Über Graphen und ihre Anwendung auf Determinantenthorie und Mengenlehre. *Math. Ann.*, **77**, 453–465.

Koppelman, D. M., & Oruc, A. Y. 1989. A self-routing permutation network. *Pages 288–295 of: Proc. Int. Conf. Para. Process.*, vol. 1. St. Charles, IL.

Koppelman, D. M., & Oruc, A. Y. 1994. The complexity of routing in Clos permutation networks. *IEEE Trans. Inform. Thy.*, **40**, 278–284.

Kubal, M. 1982. Comments on decomposition of permutation networks. *IEEE Trans. Comput.*, **31**, 265.

Lang, T. 1976. Interconnections between processors and memory modules using the shuffle-exchange network. *IEEE Trans. Comput.*, **25**, 496–503.

Lea, C.-T. 1991. Multi-$\log_2 N$ networks and their applications in high-speed electronic and photonic switching systems. *IEEE Trans. Commun.*, **38**, 1740–1749.

Lea, C.-T., & Shyy, D.-J. 1991. Tradeoff of horizontal decomposition versus vertical stacking in rearrangeable nonblocking networks. *IEEE Trans. Commun.*, **39**, 899–904.

Lee, H. Y., Hwang, F. K., & Capinelli, J. 1997. A new decomposition algorithm for rearrangeable Clos interconnection networks. *IEEE Trans. Commun.*, **44**, 1572–1578.

Lee, K. Y. 1985. On the rearrangeability of a $(2 \log N - 1)$ stage permutation network. *IEEE Trans. Comput.*, **34**, 412–425.

Lee, S., & Lu, M. 1994. Self-routing permutation networks. *IEEE Trans. Comput.*, **43**, 1319–1323.

Lenfant, J. 1978. Parallel permutations of data: A Beneš network control algorithm for frequently used permutations. *IEEE Trans. Comput.*, **27**, 637–647.

Lev, G. F., Pippenger, N., & Valiant, L. G. 1981. A fast parallel algorithm for routing in permutation networks. *IEEE Trans. Comput.*, **30**, 93–100.

Linial, N., & Tarsi, M. 1989. Interpolation between bases and the shuffle exchange network. *Euro. J. Combin.*, **10**, 29–39.

Minc, H. 1963. Upper bounds for permanents of (0,1)-matrices. *Bull. Amer. Math. Soc.*, **69**, 789–791.

Nassimi, D., & Sahni, S. 1980. A self routing Beneš network and parallel permutation networks. *IEEE Trans. Comput.*, **30**, 332–340.

Nassimi, D., & Sahni, S. 1982. Parallel algorithms to set up the Beneš permutation network. *IEEE Trans. Comput.*, **31**, 148–154.

Opferman, D., & TsaoWu, N. 1971. On a class of rearrangeable switching networks. *Bell Syst. Tech. J.*, **50**, 1579–1618.

Parker, D. S. 1980. Notes on shuffle/exchange-type networks. *IEEE Trans. Comput.*, **29**, 213–222.

Paull, M. C. 1962. Reswitching of connection networks. *Bell Syst. Tech. J.*, **41**, 833–855.

Peace, M. C. 1977. The indirect binary n-cube microprocessor array. *IEEE Trans. Comput.*, **26**, 458–473.

Pippenger, N. 1978. On rearrangeable and non-blocking switching networks. *J. Comput. Syst. Sci.*, **17**, 145–162.

Pippenger, N. 1980. A new lower bound for the number of switches in rearrangeable networks. *SIAM J. Alg. Disc. Methods*, **1**, 164–167.

Pippenger, N., & Yao, A. C. 1982. Rearrangeable networks with limited depth. *SIAM J. Alg. Disc. Methods*, **1**, 411–417.

Raghavendra, C. S. 1994. On the rearrangeability conjecture of $(2 \log_2 N - 1)$-stage shuffle/exchange network, position paper. *Comput. Arch. Tech. Com. Newsletter*, 10–12.

Raghavendra, C. S., & Boppana, R. V. 1991. On self-routing in Beneš and shuffle-exchange networks. *IEEE Trans. Comput.*, **40**, 1057–1064.

Raghavendra, C. S., & Varma, A. 1987. Rearrangeability of 5-stage shuffle/exchange network for $N = 8$. *IEEE Trans. Commun.*, **35**, 808–812.

Ramanujan, H. R. 1973. Decomposition of permutation networks. *IEEE Trans. Comput.*, **22**, 639–643.

Shannon, C. E. 1950. Memory requirements in a telephone exchange. *Bell Syst. Tech. J.*, **29**, 343–349.

Slepian, D. 1952. Two theorems on a particular crossbar switching networks. Unpublished BTL memo.

Spanke, R. A., & Beneš, V. E. 1987. N-stage planar optical permutation network. *Appl. Opt.*, **26**, 1226–1229.

Stone, H. S. 1971. Parallel processing with the perfect shuffle. *IEEE Trans. Comput.*, **20**, 153–161.

TsaoWu, N. 1974. On Neiman's algorithm for the control of rearrangeable switching networks. *IEEE Trans. Commun.*, **22**, 737–742.

Varma, A., & Raghavendra, C. S. 1988. Rearrangeability of multistage shuffle/exchange networks. *IEEE Trans. Commun.*, **36**, 1138–1147.

Waksman, A. 1968. A permutation network. *J. ACM.*, **15**, 159–163.

Wu, C.-L., & Feng, T.-Y. 1981. The universality of the shuffle-exchange network. *IEEE Trans. Comput.*, **30**, 324–332.

Chapter 4. Multicast Traffic

In the multicast traffic, an input can appear in a request more than once. If the appearance is restricted to at most f times, the traffic is called an f-cast traffic. If $f = N$, i.e., there is no constraint on f, the traffic is called *broadcast*. A switch is said to have the *fan-out* capacity if the switch itself can route multicast traffic without blocking, i.e., any inlet can be connected to any number of idle outlets regardless of other connections. In general, we assume the crossbar does have the fan-out capacity. Sometimes, we want to restrict crossbars in a given stage to perform only point-to-point connections. Then we say the stage has *no fan-out capacity*. Note that if this fan-out deficiency is due to the crossbar structure, then it is an SNB result; if due to the routing rule of an algorithm, then it is a WSNB result.

4.1 Strictly Nonblocking Multicast Networks

Bassalygo–Pinsker (1980) proved that the cost of a multicast SNB network is lower bounded by $\Omega(N^2)$. Note that the network doesn't even have to be a MIN. Here we adopt a proof given by Feldman–Friedman–Pippenger (1986).

Theorem 4.1.1. $\Omega(N^2)$ *is a lower bound of a multicast SNB network.*

Proof. Consider the graph version of a SNB network. Let i denote an input node. Let OA denote the set of output nodes i has an arc to, and let OB denote the set of other output nodes. Let LA denote the set of link-nodes i has an arc to. We shall prove $|LA| \geq |OB|$. Hence i has $|OA| + |LA| \geq |OA| + |OB| = N$ arcs. Since there are N input nodes, the graph has at least N^2 arcs.

We may assume that each link-node has at least two in-arcs for otherwise it can be deleted (and its out-arcs transferred to its predecessor) without affecting network connectivity. Let $M(i)$ denote a maximal set of node-disjoint paths from LA to OB, and let LA' denote the subset of LA appearing in $M(i)$.

Since each node v in LA' has two in-arcs, there exists a set of node-disjoint trees whose roots are input nodes other than i, and whose leaves are LA' (see Fig. 4.1.1). These trees, together with $M(i)$, define a state S in which i is

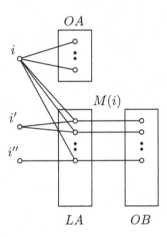

Figure 4.1.1: i has N paths

idle. Suppose there exists an output node $o \in OB$ but not in $M(i)$. Since the network is SNB, a path (i, o) node-disjoint from all paths in S can be added to S, which contradicts the maximality of $M(i)$.   □

Since the crossbar is a SNB multicast network with cost N^2, cost-wise, there is little reason to look further. However, the cost of a SNB f-cast network can be significantly lower for small f.

Most of the known results on SNB multicast networks are for 3-stage clos networks. Four models have been studied in the literature:

Model 0. No restriction on fan-out capability (since this is the standard model, sometime its specification is omitted).

Model 1. Stage 1 has no fan-out capability.

Model 2. Stage 2 has no fan-out capability.

Model 3. Stage 3 has no fan-out capability.

Each model has been studied for both the open-end and the closed-end traffic. We first give some fundamental relation between the two types of traffic. We simplify a proof of Hwang and Liaw (2000) to obtain

Lemma 4.1.2. *A 3-stage Clos network is not f-cast SNB for the open-end traffic if it is not so for closed-end traffic.*

Proof. Consider a sequence S of closed-end traffic leading to the blocking of a request c. Then S is also a sequence of open-ended traffic. Further, any routing of c in the open-end environment is also a routing in the closed-end environment. Hence the blocking of c in the latter implies the same in the former. □

The other direction does not hold in general. But,

Lemma 4.1.3. *If the first two stages of a 3-stage Clos network have the fan-out capability, then it is f-cast SNB for the open-end traffic if and only if it is for the closed-end traffic.*

Proof. By Lemma 4.1.2, it suffices to prove for the "only if" part. Consider a request c in the open-end environment where the input i of c is already connected to some other outputs. Let $C(i)$ denote the set of existing connections involving i. By assumption, c can be routed in a path p if $C(i)$ is ignored. But the presence of $C(i)$ does not block the route p since c can share links with $C(i)$. So if c is blocked in the open-end traffic, it must be because c is not allowed to use the route p. This can happen only if either the first stage or the second stage has no fan-out capability. In the former case, c must go to the middle switch where $C(i)$ went, which may be different from the route p. In the latter case, after c follows the route p to a middle switch, it must follow $c' \in C(i)$ to an output switch which, again, maybe different from the route p. □

Attempts to give necessary and/or sufficient conditions for SNB under model 0 started earlier. Masson-Jordan (1972) give a sufficient condition which was later found to be a WSNB result since the routing follows a certain rule. Giacomazzi-Trecordi (1995) gave necessary and sufficient condition, while Hwang-Liaw (2000) gave different and more explicit conditions. Pattavina-Tesei (2002) gave a counterexample to Giacomazzi-Trecordi's condition, and extended the condition of Hwang-Liaw to the case when a multicast call must involve at least f' outputs. Hwang-Liaw also gave necessary and sufficient conditions for models 1, 2 and 3 under both closed and open traffic. Recently, Hwang (2002) umfied and simplified the arguments of Hwang-Liaw to derive necessary and sufficient condition for all these models plus some WSNB models.

Suppose the current request c is from an input i on the input switch I to k outputs on a set O of output switches. Let $b(I)$ denote the potential maximum number of middle switches occupied by paths from I which c cannot use, while ignoring the boundary effect from the output side. Then the actual maximum number is

$$\min\{b(I), N_2 - k\} \tag{4.1.1}$$

since each path through these middle switches must end at a distinct output, but there are only $N_2 - k$ of them available.

Let $O_j \in O$ be an output switch containing k_j outputs in c. Then each of the other $n_2 - k_j$ outputs can occupy a distinct middle switch as long as there are enough inputs to generate requests involving these $n_2 - k_j$ outputs. Since paths generated by inputs of I are already counted in (4.1.1), there are $N_1 - n_1$ inputs available. Let $b(N_1)$ denote the potential maximum number of middle switches the paths from these $N_1 - n_1$ inputs can occupy while ignoring the output side.

Another point to consider is that some of the $n_2 - k_j$ outlets may already be used in (4.1.1). There are $N_2 - k - (n_2 - k_j)$ outlets not in the k-request and not from O_j. So if

$$N_2 - k - n_2 + k_j < b(I) < N_2 - k,$$

then

$$b(I) - (N_2 - k - n_2 + k_j)$$

among the $n_2 - k_j$ outputs from O_j are already consumed in (4.1.1). Hence only $(N_2 - k - b(I))^+$ outputs from O_j can generate new paths, where $(x)^+ = \max\{x, 0\}$. Summarizing, the maximum number of new paths generated by available outputs from O_j is

$$\min\{n_2 - k_j, b(N_1), (N_2 - k - b(I))^+\} \tag{4.1.2}$$

Suppose c involves outputs on several output switches. In general, it suffices to consider just one of them since they do not compete for links to the middle stage and the condition derived for one is the same for the other. An exception is when those output switches involved in c are required to be routed through the same middle switch, as in model 1, then we must consider the output switches together because a failure for one is a failure for all.

Let $b(c)$ denote the minimum number of middle switches guaranteeing to route c when there is no other connection. First consider the case that the routing of c to the output switches are independent. Then to route c among other connections, we must have $b(c)$ additional switches not counted in (4.1.1) and (4.1.2). In the worst-case scenario where the switches in (4.1.1) and (4.1.2) are all distinct, then a sufficient condition to route c is

$$m \geq \min\{b(I), N_2 - k\} + \min\{n_2 - k_j, b(N_1), (N_2 - k - b(I))^+\} + b(c)$$
$$= \min\{b(I) + n_2 - k_j + b(c), b(I) + b(N_1) + b(c), N_2 - k + b(c)\},$$
$$\text{(4.1.3)}$$

obtained by pairing $b(I)$ with each term in the second min. Note that if the first min is $N_2 - k$, then the second min must be $(N_2 - k - b(I))^+ = 0$, and the condition becomes $N_2 - k + b(c)$ which is the last term in (4.1.3). A careful examination of the above arguments reveals that the worst-case scenario generating (4.1.1) and (4.1.2) can happen, hence (4.1.3) is also a necessary condition.

To save space, let m^0 denote the number such that $m \geq m^0$ is necessary and sufficient for $C(n_1, r_1, n_2, r_2, m)$ to be f-cast SNB(WSNB for the no-split model).

Lemma 4.1.4. *Suppose the routings of c to the output switches in O are independent. Then*

$$m^o = \max_{k_j, k} \min\{b(I) + n_2 - k_j + b(c), b(I) + b(N_1) + b(c), N_2 - k + b(c)\}.$$

Corollary 4.1.5. *If $b(c)$ is independent of k_j, then*

$$m^o = \max_k \min\{b(I) + n_2 - 1 + b(c), b(I) + b(N_1) + b(c), N_2 - k + b(c)\}.$$

Lemma 4.1.4 applies to models 0,2,3, while Corollary 4.1.5 to any model whose output stage has fan-out capability.

Theorem 4.1.6. *For model 0 under the closed-end traffic,*

$$m^o = \min\{(n_1 - 1)f + n_2, (N_1 - 1)f + 1, N_2\}.$$

Proof. $b(I) = (n_1 - 1)f$, $b(N_1) = (N_1 - n_1)f$ and $b(c) = 1$. By Corollary 4.1.5,

$$m = \max_k \min\{(n_1 - 1)f + n_2, (N_1 - 1)f + 1, N_2 - k + 1\}$$
$$= \min\{(n_1 - 1)f + n_2, (N_1 - 1)f + 1, N_2\} \text{ at } k = 1.$$

$\square$

Patavina and Tesei (2002) extended Theorem 4.1.6 to the case that an output can also generate a multicast call up to f_1 inputs.

Corollary 4.1.7. *m^o is the same for model 0 under the open-end traffic.*

Proof. By Lemma 4.1.3, we only need to prove that if m^o suffices for the closed-end traffic, then it suffices for the open-end traffic.

Consider a k-request from i such that i has already been connected to f' outputs, $k + f' \leq f$. Transform the open-end traffic to closed-end by combining all requests from the same input into one (keeping their paths intact). In particular, the connections of i to the f' outputs are deleted and the corresponding requests join the k-request as the current request. Then this $(k + f')$-request is routable since m^o suffices for closed-end traffic.

Back to the open-end traffic, we route the current k-request using the same paths as in the transformed traffic. Then the only possible overlapping of these paths is with the existing paths from i, which is alright since in model 0, paths from the same input are allowed to overlap. $\square$

Theorem 4.1.8. *For model 2 under the closed-end traffic,*

$$m^o = \min\{(n_1 - 1)f + n_2 - 1 + \min\{f, r_2\}, (N_1 - 1)f + \min\{f, r_2\}, N_2\}.$$

Proof. $b(I) = (n_1 - 1)f$, $b(N_1) = (N_1 - n_1)f$, and $b(c) = \min\{k, r_2\}$ since if $f > r_2$, r_2 additional middle switches still suffice by routing each such middle switch to a distinct output switch, and then use the fan-out capability of the output switches to reach multiple outputs. The reason that f, but not $\min\{f, r_2\}$, is in $b(I)$ and $b(N_1)$ is because under the SNB rule, each connection can be routed arbitrarily, as long as the paths are available. Hence a connection can use two different paths to reach two inputs on the same output switch, even if it is a waste. By Corollary 4.1.5,

$$m^o = \max_k \min\{(n_1 - 1)f + n_2 - 1 + \min\{k, r_2\}, (N_1 - 1)f + \min\{k, r_2\},$$
$$N_2 - k + \min\{k, r_2\}\}$$
$$= \min\{(n_1 - 1)f + n_2 - 1 + \min\{f, r_2\}, (N_1 - 1)f + \min\{f, r_2\}, N_2\}$$
$$\text{at } k = \min\{f, r_2\}.$$

$\square$

Theorem 4.1.9. *For model 2 under the open-end traffic,*

$$m^o = \min\{n_1 f + n_2 - 1, N_1 f, N_2\}.$$

Proof. $b(I) = n_1 f - k$, $b(N_1)$ and $b(c)$ are same as the closed-end case. By Corollary 4.1.5,

$$m^o = \max_f \min\{n_1 f - k + n_2 - 1 + \min\{k, r_2\}, N_1 f - k + \min\{k, r_2\},$$
$$N_2 - k + \min\{k, r_2\}\}$$
$$= \min\{n_1 f + n_2 - 1, N_1 f, N_2\} \text{ at any } k \leq r_2.$$

$\square$

Note that the conditions for the closed-end traffic and the open-end traffic are different.

Theorem 4.1.10. *For model 3 under the closed-end traffic,*

$$m^o = \min\{(n_1 - 1)f + n_2, (N_1 - 1)f + \min\{f, n_2\}, N_2\}.$$

Proof. It is easily argued that $b(I) = (n_1 - 1)f$ and $b(N_1) = (N_1 - n_1)f$. To derive $b(c)$, note that if c involves outputs on the same output switch, then these outputs must each be routed through a distinct middle switch. On the other hand, outputs of different output switches can be routed through the same middle switch. Then $b(c) = k_j$, where $k_j \leq \min\{k, n_2\}$. By Lemma 4.1.4,

$$m^o = \max_{k, k_j} \min\{(n_1 - 1)f + n_2 - k_j + k_j, (N_1 - 1)f + k_j, N_2 - k + k_j\}$$
$$= \max_k \min\{(n_1 - 1)f + n_2, (N_1 - 1)f + \min\{k, n_2\}, N_2 - k + \min\{k, n_2\}\}$$
$$\text{at } k_j = \min\{k, n_2\}$$
$$= \min\{(n_1 - 1)f + n_2, (N_1 - 1)f + \min\{f, n_2\}, N_2\} \text{ at } k = \min\{f, n_2\}.$$

$\square$

Corollary 4.1.11. m^o *is the same for model 3 under the open-end traffic.*

Proof. Again, we transform the open-end traffic to closed-end as in the proof of Corollary 4.1.7. Note that in the transformed traffic, paths from i to outputs in the same output switch must be routed through different middle switches. Further, these paths can be interchanged (at the input switch) without affecting the routability. In routing the current request in the transformed traffic, make all such necessary changes such that no path from i to $o \in O_j$ goes through a middle switch which routes i to $o' \in O_j$ in an existing connection in the open-end traffic. Then we can route the current request in the open-end traffic using the same paths as in the transformed traffic. $\square$

Finally, we deal with model 1. In this model, we need to find one middle switch which can route to all output switches in O. Therefore, we must replace $n_2 - k_j$ in (4.1.2) by $|O|n_2 - k$. The corresponding change in m^0 is

$$m^o = \max_{k,|O|} \min\{b(I)+|O|n_2-k+b(c), b(I)+b(N_1)+b(c), N_2-k+b(c)\}. \quad (4.1.4)$$

Theorem 4.1.12. *For model 1 under the closed-end traffic, $N_2 \geq n_1$, $n_2 \geq 2$,*

$$m^o = \min\{N_1, N_2 - \lceil (N_2 + 1 - n_1)/n_2 \rceil + 1\}.$$

Proof. Since each k-request must go to a single middle switch, $b(I) = n_1 - 1$, $b(N_1) = N_1 - n_1$ and $b(c) = 1$. By (4.1.4),

$$\begin{aligned} m^o &= \max_{k,|O|} \min\{n_1 + |O|n_2 - k, N_1, N_2 - k + 1\} \\ &= \max_{k} \min\{n_1 + (n_2 - 1)k, N_1, N_2 - k + 1\} \ at \ |O| = k. \end{aligned} \quad (4.1.5)$$

Since the first term is increasing in k, and the third decreasing, the maximum should occur at

$$n_1 + (n_2 - 1)k = N_2 - k + 1,$$

which is solved by $k^o = (N_2 + 1 - n_1)/n_2$ if the integrability of k is ignored. It is easily verified that the maximum over k is either at $\lceil k^o \rceil$ or $\lfloor k^o \rfloor$. Further, since, $N_2 - \lceil k^o \rceil + 1 \geq n_1 + (n_2 - 1)\lfloor k^o \rfloor$, we obtain Theorem 4.1.12. $\square$

For open-end traffic, Hwang-Liaw gave the following example (see Fig. 4.1.2) to show that no m^o is large enough to guarantee SNB (i_1 cannot be connected to o_2).

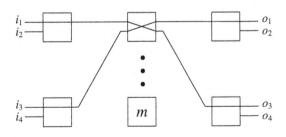

Figure 4.1.2: (i_1, o_2) cannot be routed

Table 4.1.1 summarizes the results of the four models for easy comparisons. Line 1 in each model represents the closed-end case and Line 2 the open-end case.

model	main condition	input boundary	output boundary
0	$(n_1 - 1)f + n_2$	$(N_1 - 1)f + 1$	N_2
	same		
1	$n_1 + (n_2 - 1)(N_2 + 1 - n_1)/n_2$	N_1	$N_2 - (N_2 + 1 - n_1)/n_2 + 1$
	blocking		
2	$(n_1 - 1)f + n_2 - 1 + \min\{f, r_2\}$	$(N_1 - 1)f + \min\{f, r_2\}$	N_2
	$n_1 f + n_2 - 1$	$N_1 f$	N_2
3	$(n_1 - 1)f + n_2$	$(N_1 - 1)f + n_2$	N_2
	same		

Table 4.1.1. Results of the 4 models

From Theorem 4.1.1 we know f must be bounded to improve over $O(N^2)$ crosspoints. It is easily verified that for $f = r_2$, all these models need $O(N^{5/3})$ crosspoints by setting $n_1 = O(N^{1/3})$ and $n_2 = O(N^{2/3})$, except model 1 needs $O(N^2)$ crosspoints. For $C(n, m, r)$, every model needs $O(N^2)$ crosspoints.

For $f = n_2$, we can improve over $O(N^{5/3})$ by recursively extending a 3-stage clos network to a $(2k + 1)$-stage network as done in Chapter 2. The only difference is that a crossbar is replaced by an $O(N^{3/2})$-crosspoints 3-stage SNB network in Chapter 2, but by an $O(N^{5/3})$-crosspoints network here. So the decreasing rate is slower.

In their work on $\log_2(N, k, m)$ as a point-to-point network, Lea also mentioned that some of the results can apply to multicast network. Kabacinski and Danilewitz (2002) gave a definite answer by proving a necessary and sufficient condition for $\log_2(N, 0, m)$ to be broadcast strictly nonblocking (see Lemma 4.1.16 below). Recently, Hwang, Wang and Tan (2003) extended the study to the f-cast case.

Consider a request from input i to output o. Then the (i, o) channel graph is simply the path from i to o containing $n + 1$ links $L_0, L_1, \ldots, L_n$. A connection from $i' \neq i$ to $o' \neq o$ is called an j-intersecting connection if it contains L_j. Note that a connection can be both j-intersecting and j'-intersecting. An input is called an j-intersecting input if it can start a j-intersecting connection. Note that a j-intersecting input is also a j'-intersecting input for $j < j' \leq n - 1$. An input is j-marginal if it is j-intersecting but not $(j - 1)$-intersecting. Similarly, an output is j-intersecting if it can end a j-intersecting connection. A j-intersecting output is also a j'-intersecting output for $1 \leq j' < j$. An output is j-marginal if it is j-intersecting but not $(j+1)$-intersecting. Define

$|I_j|$= number of j-intersecting inputs,
$|I'_j|$= number of j-marginal inputs,
$|O_j|$= number of j-intersecting outputs,
$|O'_j|$= number of j-marginal outputs,
$|P_j|$= number of j-intersecting, but not j'-intersecting for $j' < j$ connections.

Then

$$|I_j| = d^j - 1 \quad |I'_j| = d^j - d^{j-1} \quad |O_j| = d^{n-j} - 1 \quad |O'_j| = d^{n-j} - d^{n-j-1}$$
$$|P_j| = \min\{|I'_j|f, |O_j|\}.$$

Lemma 4.1.13. $|P_j| = |I'_j|f$ if $f < d^{n-2j}$.

Proof. $|I'_j|f - |O'_j| = (d^j - d^{j-1})f - (d^{n-j} - d^{n-j-1}) \geq 0$ if and only if $f \geq d^{n-2j}$. Since $|O'_j| \leq |O_j|$, Lemma 4.1.13 follows. □

Lemma 4.1.14. If $f \geq d^{n-2j}$, then $(|I'_j| + |I'_{j+1}|)f \geq |O_j|$.

Proof.

$$\begin{aligned}
(|I'_j| + |I'_{j+1}|)f - |O_j| &= (d^{j+1} - d^{j-1})f - (d^{n-j} - 1) \\
&\geq (d^{j+1} - d^{j-1})d^{n-2j} - d^{n-j} + 1 \\
&\geq d^{n-j-1} + 1 > 0 .
\end{aligned}$$

□

Theorem 4.1.15. $Log_d(N, 0, p)$ *is f-cast strictly nonblocking if and only if* $p > (d^j - 1)f + d^{n-j-1} - 1$ *for* $d^{n-2j} > f \geq d^{n-2j-2}, j = 0, 1, \ldots, \lfloor \frac{n-1}{2} \rfloor$.

Proof. Suppose current request involves x outputs. Since we can connect the x outputs independently one by one, we may assume $x = 1$. Let the request be (i, o). Suppose $d^{n-2j} > f \geq d^{n-2j-2}$ for some fixed j. Then the upper bound implies $d^{n-2k} > f$ for all $0 \leq k \leq j$. By Lemma 4.1.13,

$$|I'_k|f < |O'_k| \qquad\qquad \text{for all } 0 \leq k \leq j. \text{ Hence}$$
$$|P_k| = |I'_k|f = (d^k - d^{k-1})f \qquad \text{for } 0 \leq k \leq j.$$

Further by Lemma 4.1.14

$$(|I'_{j+1}| + |I'_{j+2}|)f \geq |O_{j+1}|,$$

which says that the $(j+1)$-marginal and $(j+2)$-marginal inputs start enough paths to occupy all remaining outputs at step $j+1$. So the total number of blocked copies is

$$\sum_{k=0}^{j} |I_k|f + |O_{j+1}|$$

$$= \sum_{k=1}^{j} (d^k - d^{k-1})f + d^{n-j-1} - 1$$

$$= (d^j - 1)f + d^{n-j-1} - 1.$$

$\square$

Set $j = 0$, we obtain

Lemma 4.1.16. $Log_d(N, 0, p)$ *is broadcast strictly nonblocking if and only if* $p > d^{n-1} - 1$.

Set $j = \lfloor (n-1)/2 \rfloor$, we obtain

Corollary 4.1.17. $Log_d(N, 0, p)$ *is point-to-point strictly nonblocking if and only if* $p > d^{\lfloor (n-1)/2 \rfloor} + d^{\lceil (n-1)/2 \rceil} - 2$.

Hwang-Wang-Tan also studied for general m.

Theorem 4.1.18. $Log_d(N, m, p)$ *is f-cast strictly nonblocking if and only if*

$$p \geq \begin{cases} \frac{N-1}{d}, & \text{if } f \geq \frac{N-d}{d-1}, \\ \frac{(d-1)(f+1)k}{d} + \frac{N-1-(f+1)(d^k-1)}{d^{k+1}}, & \text{if } \frac{N-d^k}{d^k-1} > f \geq \frac{N-d^{k+1}}{d^{k+1}-1}, \\ & \quad 1 \leq k \leq m-1, \\ \frac{(d-1)(f+1)m}{d} + d^{n-m-1} - 1, & \text{if } \frac{N-d^m}{d^m-1} > f \geq d^{n-m-2}, \\ \frac{(d-1)(f+1)m}{d} + (d^j-1)f + d^{n-m-j-1} - 1, & \text{if } d^{n-m-j} > f \geq d^{n-m-j-1}, \\ & \quad 1 \leq j \leq \lfloor (n-m-1)/2 \rfloor. \end{cases}$$

Proof. Again, we may assume that the current request is point-to-point and from input i to output o. Note that the channel graph has d^j stage-j links. Hence a j-intersecting connection blocks only $1/d^j$ copy of the channel graph.

Since a j-intersecting connection and a $(n-j)$-intersecting connection blocks the same fraction of a copy, we count them together. Furthermore, for $1 \leq j \leq m$, a j-intersecting connection can reach all outputs, and a $(n-j)$-intersecting connection all inputs. Thus we can always arrange them such that the inputs and the outputs are all distinct, as long as there are enough outputs. For $j \geq m$ we modify the definition of $|P_j|$ to be the number of j-intersecting, but not j'-intersecting for $j' > j$, connections. Then

$$|P_1| + |P_{n+m-1}| = \min\{(d-1)f + d - 1, N - 1\} = \min\{(d-1)(f+1), N - 1\}.$$

(i) $f \geq (N - d)/(d - 1)$

Then $|P_1| + |P_{n-1}| = N - 1$. Since each 1-intersecting connection blocks $1/d$ copy, the total number of copies blocked is $(N - 1)/d$.

(ii) $(N - d^j)/(d^j - 1) > f \geq (N - d^{j+1})/(d^{j+1} - 1), 1 \leq k \leq m - 1$.

The upper bound is equivalent to

$$|P_k| + |P_{n-k}| = (d^k - 1)(f + 1) < N - 1$$

while the lower bound is equivalent to

$$|P_{k+1}| + |P_{n-k-1}| = (d^{k+1} - 1)(f + 1) \geq N - 1.$$

Therefore the number of copies blocked is

$$\sum_{j=1}^{k} \frac{(d^j - 1)(f + 1) - (d^{j-1} - 1)(f + 1)}{d^j} + \frac{N - 1 - (d^k - 1)(f + 1)}{d^{k+1}}$$

$$= \frac{(d - 1)(f + 1)k}{d} + \frac{N - 1 - (d^k - 1)(f + 1)}{d^{k+1}}.$$

(iii) $(N - d^m)/(d^m - 1) > f$

We need to count further blocking caused by the middle $n - m$ stages. The network is reduced to d^m copies of $BY^{-1}(n - m, 0)$. In each copy, the (i, o) channel graph has a unique path going through an input i' and an output o' of $BY^{-1}(n - m, 0)$. Let $P'(j)$ denote the set of path intersecting the (i', o') path at the jth link in a $BY_d^{-1}(n - m, 0)$. Then $|P'(j)| = d^{m+j} - d^m = d^m(d^j - 1)$, which is d^m times of $|P(j)|$ of a normal $BY^{-1}(n - m, 0)$. However, to block a copy of $BY^{-1}(n - m, 0)$ needs to block d^m copies of $BY^{-1}(n - m, 0)$. Hence the net effect of $P'(j)$ is same as $P(j)$ of a normal $BY^{-1}(n - m, 0)$. We can apply Theorem 4.1.15 with n replaced by $n - m$ everywhere.

$\square$

Set $f = d^n - 1$, we obtain

Corollary 4.1.19. $Log_d(N, m, p)$ *is broadcast strictly nonblocking if and only if* $p \geq N/d = d^{n-1}$.

It is surprising that the condition on p is independent of m.

Corollary 4.1.20. $Log_d(N, m, p)$ *is point-to-point strictly nonblocking if and only if*

$$p \geq \frac{2(d-1)m}{d} + d^{\lfloor \frac{n-m-1}{2} \rfloor} + d^{\lceil \frac{n-m-1}{2} \rceil} - 1.$$

Proof. Set $j = \lfloor (n - m - 1)/2 \rfloor$ in Theorem 4.1.18. □

4.2 Wide-Sense Nonblocking Multicast Networks

In Chapter 2 we found that there is not only a paucity of WSNB results, but also a paucity of routing algorithms since there are not many factors which can be used to define routing algorithms. For multicast networks, we have at least one more dimension, the degree of broadcasting, to play with. Consequently, we have more meaningful routing algorithms to discuss.

The simplest control is the 0-1 control, i.e., whether a stage can broadcast (fan-out, if dealing with crossbars). The first 0-1 algorithm, also the first multicast WSNB algorithm, is the no-split algorithm of Masson-Jordan on 3-stage clos network. This algorithm specifies that outputs in the same output switch in a multicast request must be connected by using the fan-out of the output crossbar (implying that these outputs must share a path until reaching the output crossbar). This rule is so natural that Masson-Jordan took it for granted and thought their sufficient conditions hold for arbitrary routing (hence SNB). Feldman-Friedman-Pippenger (1986) noted the implicit no-split rule and correctly claimed the sufficient conditions are for WSNB. Hwang (2002) used the unifying approach of the last section to give a necessary and sufficient condition under the no-split rule.

Theorem 4.2.1. *For the no-split algorithm under the closed-end traffic,*

$$m^o = \min\{(n_1 - 1) \min\{f, r_2\} + n_2, (N_1 - 1) \min\{f, r_2\} + 1, N_2\}.$$

Proof. A k-request can use at most $\min\{k, r_2\}$ paths to reach the output stage. Hence $b(I) = (n_1 - 1) \min\{f, r_2\}$ and $b(N_1) = (N_1 - n_1) \min\{f, r_2\}$, while

$b(c) = 1$. By Corollary 4.1.5,

$$m^0 = \max_k \min\{(n_1 - 1)\min\{f, r_2\} + n_2, (N_1 - 1)\min\{f, r_2\} + 1, N_2 - k + 1\}$$
$$= \min\{(n_1 - 1)\min\{f, r_2\} + n_2, (N_1 - 1)\min\{f, r_2\} + 1, N_2\} \text{ at } k = 1.$$

$\square$

Corollary 4.2.2. *The open-end traffic has the same m^0.*

Proof. If the current request from i contains an output o from the same output switch O_j through which an existing path from i to O_j was routed, then use the same path to route (i, o). So we need only be concerned with the part of the current request whose outputs are from output switches not connected to i in the existing traffic. Since this part of the current request is allowed to overlap with existing connections from input i, we can use the same routing method as in model 0. $\square$

Other algorithms in the 0-1 class are those corresponding with models 1, 2, 3 except the fact that fan-out is not needed in a given stage is treated as an algorithm rule.

A more effective control is on the amount of broadcasting. Although the root of such an algorithm was seeded in a multicast RNB algorithm of Kirkpatrick-Klawe-Pippenger (1988) (which was observed by Hwang (1998) to be really a WSNB algorithm), Yang-Masson (1990) first explicitly suggested a multicast WSNB algorithm of this category. The following lemma is crucial to their results (the simplifying proof is due to H.G.Yeh).

Lemma 4.2.3. *Consider s subsets of $\{1, \ldots, r\}$ such that each element can appear in at most n subsets, $n \leq s$. Suppose the intersection of every x subset is nonempty for some $1 \leq x \leq r$. Then $s \leq nr^{1/x}$.*

Proof. Consider the $r \times m$ binary matrix B where rows are elements, columns are subsets, and cell $b_{ij} = 1$ if subset j contains element i. Let X be a set of x columns. Then the intersection of X is nonempty if and only if there exists a row intersecting every column in X. On the other hand, each row has weight n, thus can yield only $\binom{n}{x}$ all-1 x-vectors. The condition of Lemma 4.2.3 translates to

$$\binom{n}{x} r \geq \binom{s}{x}, \quad \text{or} \quad n(n-1)\cdots(n-x+1)r \geq s(s-1)\cdots(s-x+1).$$

Since $n \leq s$ implies $(n-i)/n \leq (s-i)/s$ for $i = 1, \ldots, x-1$, we have

$$n^x \frac{n-1}{n} \cdots \frac{n-x+1}{n} r \geq s^x \frac{s-1}{s} \cdots \frac{s-x+1}{s}$$

implies $n^x r \geq s^x$, or $s \leq n r^{1/x}$.   □

Theorem 4.2.4. *Under the routing algorithm that any request can use at most p middle crossbars, $C(n_1, r_1, m, n_2, r_2)$ is WSNB for closed-end f-cast traffic, $f \leq r_2$, if $m > (n_1 - 1)p + (n_2 - 1)f^{1/p}$.*

Proof. Consider an f_2-request. Each of the $n_1 - 1$ co-inputs can engage at most $(n_1 - 1)p$ middle crossbars due to the routing algorithm. There are at most f_2 output switches each containing $n_2 - 1$ co-outputs. Since $m - (n_1 - 1)p > (n_2 - 1)f_2^{1/p}$, by Lemma 4.2.3, there exists a set of p middle crossbars with empty intersection with respect to the f_2 output crossbars. In other words, for each such output crossbar Z, there exists a middle crossbar among the p crossbars which does not carry a connection going to Z and hence can route the request to Z.   □

Setting $p=1$ is equivalent to the condition that input switches have no fan-out capability the result is comparable to Theorem 4.1.12 except the boundary condition is not considered. Setting $p = n_2$ is equivalent to what Kirkpatrick, Klawe and Pippenger did in Lemma4.3.4.

Corollary 4.2.5. $m = O(n \log f_2 / \log \log f_2)$ *suffices.*

Proof. Set $p = \log_2 f_2 / 2 \log_2 \log_2 f_2$.   □

By setting $n_1 = n_2 = O(N^{1/2})$ and $f_2 = r_2$ (using the no-split rule), the cost of the Yang–Masson network is $O(N^{3/2} \log N / \log \log N)$.

Yang–Masson also gave the following $O(n_2 f_2)$ time algorithm to find the set P of p middle crossbars which can route the new request.

Suppose the new request engages a set L of f output crossbars. Let M denote the set of middle crossbars not engaged by the co-inputs. Suppose $|M| = (n_2 - 1)w$ for some $w > f^{1/p}$. Let the ith member of M be represented by the subset $L_i \in L$ of output crossbars to which it carries a connection. Without loss of generality, assume $|L_1| = \min |L_i|$. Set $L_1^1 = L_1$. Since M can engage at most $(n_2 - 1)f$ co-outputs,

$$|L_1^1| \leq \frac{(n_2 - 1)f}{|M|} = \frac{(n_2 - 1)f}{(n_2 - 1)w} = \frac{f}{w}.$$

Define $L_i^2 = L_i \cap L_1$. Without loss of generality, assume $|L_2^2| = \min |L_i^2|$. Then

$$|L_2^2| \le \frac{(n_2 - 1)f/w}{(n_2 - 1)w} = \frac{f}{w^2} \,.$$

Repeat this process. After $\ell = \lfloor \log_w f \rfloor$ iterations, the minimum intersection set is empty. $P = \{L_1^1, L_2^2, \ldots, L_\ell^\ell\}$. Note that

$$\ell \le \log_w f = \frac{\log f}{\log w} < \frac{\log f}{\frac{1}{p}\log f} = p\,.$$

The time complexity of this algorithm is

$$|M|\left(f + \frac{f}{w} + \frac{f}{w^2} + \cdots + \frac{f}{w^{\log f_2}}\right) \le \frac{w}{w-1}|M|f\,.$$

In most cases $\frac{w}{w-1}$ is a small constant. Then the time complexity is just $O(|M|f) \le O(|M|f_2)$, which is $O(n_2)$ when p is set to $\log f_2$ and f_2 is treated as a constant.

Consider the routing of a frame. Since an f-request takes $O(n_2 f)$ time, and the sum of f over all requests is upper bounded by N_2, the time complexity of routing a frame is $O(n_2 N_2)$.

In a subsequent paper, Yang–Masson (1996) showed that $m \ge O(N \log r / \log \log r)$ is also necessary for WSNB under the following three algorithms:

(i) Save the unused.

(ii) Select the minimum number of middle crossbars to connect a request.

(iii) Select sequentially the middle crossbar which can carry the largest number of outputs in the current request.

Yang–Masson's results on 3-stage Clos network can be extended to $2k + 1$ stages. They proved

Theorem 4.2.6. *The cost of the Yang–Masson WSNB ($2k + 1$)-stage Clos network with closed-end traffic is*

$$G_{2k+1}(N) \le O(N^{1+\frac{1}{k+1}}(\log N/\log \log N)^{\frac{k+2}{2}-\frac{1}{k+1}})\,.$$

Proof. By Corollary 4.2.5 (setting $f_2 = r$),

$$G_3(N) = 2rnm + mr^2 \le O(nr(\log r/\log \log r)(n+r)) = O(N^{3/2}\log N/\log \log N)$$

by setting $n = r = \sqrt{N}$. Therefore Theorem 4.2.6 holds for $k = 1$. The general k case is proved by induction.

$$
\begin{aligned}
G_{2k+1}(N) &\leq m\left(2nr + G_{2k-1}(r)\right) \leq O\left(\frac{N\log N}{\log\log N}\left(\frac{N}{r} + \frac{G_{2k-1}(r)}{r}\right)\right) \\
&= O\left(\frac{N\log r}{\log\log r}\left(\frac{N}{r} + r^{\frac{1}{k}}(\log r/\log\log r)^{\frac{k+1}{2}-\frac{1}{k}}\right)\right).
\end{aligned}
$$

Set

$$
r = \frac{N^{\frac{k}{k+1}}}{(\log N/\log\log N)^{\frac{k}{2}-\frac{1}{k+1}}}.
$$

Then

$$
\frac{\log r}{\log\log r} = \frac{k}{k+1}\frac{\log N}{\log\log N}\alpha_N, \text{ where } \lim_{N\to\infty}\alpha_N \to 1.
$$

Hence

$$
\begin{aligned}
G_{2k+1}(N) &\leq O\left(\frac{N\log N}{\log\log N}\left(N^{\frac{1}{k+1}}\left(\frac{\log N}{\log\log N}\right)^{\frac{k}{2}-\frac{1}{k+1}}\right.\right. \\
&\quad \left.\left. + N^{\frac{1}{k+1}}\left(\frac{\log N}{\log\log N}\right)^{-\frac{1}{2}+\frac{1}{k(k+1)}}\left(\frac{\log N}{\log\log N}\right)^{\frac{k+1}{2}-\frac{1}{k}}\right)\right) \\
&= O\left(N^{1+\frac{1}{k+1}}\left(\frac{\log N}{\log\log N}\right)^{\frac{k+2}{2}-\frac{1}{k+1}}\right).
\end{aligned}
$$

$\square$

Feldman, Friedman and Pippenger (1988) proposed a 2-stage network where the first stage consists of concentrators and the second stage crossbars. All broadcasting are done inside the crossbar. Consider a request from input i. The algorithm chooses an output of the concentrator to connect to i such that the number of busy neighbors (outputs) of each input is upper bounded by a prespecified parameter. The concentrator is constructed by finite geometry.

For q a prime power, let $G(q)$ be a (q^3, q^2) bipartite graph where inputs are labeled by (a, b, c), outputs by (x, y), $a, b, c, x, y \in GF(q)$, and an edge exists between (a, b, c) and (x, y) if and only if $ax^2 + bx + c = y$. Clearly, each input vertex has q edges, one for each choice of x. Since

$$
ax^2 + bx + c - (a'x^2 + b'x + c') = 0
$$

has at most two solutions of x, any two input vertices share at most two neighbors.

A switch or a network is called a *concentrator* if any k inputs can be simultaneously connected to some k outputs. Normally, it is used in RNB networks. By allowing the existence of some previous connections, the notion concentrator can be extended to a WSNB environment. Thus an F-limited concentrator is a network such that if no more than F outputs are already connected, then a set of k inputs can always be connected to some k outputs. If there is an upper bound f of k, then it is a (F, f)-limited concentrator.

Lemma 4.2.7. $G(5^\lambda)$ *is a WSNB* $(2 \cdot 5^{2\lambda-2})$-*limited concentrator with closed-end traffic.*

Proof. Let the safe states be those in which every input vertex has at most $3 \cdot 5^{\lambda-1}$ neighbors connected to other inputs. It is easily verified that the empty state and the states below a safe state are safe. Let a request be represented by its input. Consider a safe state s where a request v cannot be routed. It suffices to prove that s has at least $2 \cdot 5^{2\lambda-2}$ busy outputs.

Since v is idle, v has at most $3 \cdot 5^{\lambda-1}$ busy neighbors. Hence v has at least

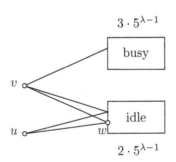

Figure 4.2.1: Critical input and output

$2 \cdot 5^{\lambda-1}$ idle neighbors w such that $s \cup \{v, w\}$ is an unsafe state. Call these w the *critical outputs*. Note that $s \cup \{v, w\}$ unsafe implies that each w has a neighbor $u \neq v$ with $3 \cdot 5^{\lambda-1}$ neighbors connected to other inputs. Call u a *critical input*. Each critical input u is adjacent to at most two critical outputs for otherwise u and v would share more than two neighbors (see Fig. 4.2.1). Thus the $2 \cdot 5^{\lambda-1}$ critical outputs are adjacent to at least $5^{\lambda-1}$ distinct critical inputs. A set of $5^{\lambda-1}$ critical inputs have at least $5^{\lambda-1}(3 \cdot 5^{\lambda-1})$ busy neighbors, not necessarily distinct. But two critical inputs can share at most two neighbors. Hence the number of distinct busy outputs is at least

$$5^{\lambda-1}(3 \cdot 5^{\lambda-1}) - \binom{5^{\lambda-1}}{2} > 2 \cdot 5^{2\lambda-2}.$$

$\square$

Note that the proof depends on the assumption that v is idle, which implies closed-end traffic.

Lemma 4.2.8. *Let H be a WSNB F-limited (n, m)-concentrator and G be a WSNB (F, f)-limited (m, ℓ)-network. Then $H \circ G$ is a WSNB (F, f)-limited f-cast (n, ℓ)-network for open-end traffic.*

Proof. Let the safe states of $H \circ G$ be those in which the induced states of H and G are safe. Clearly, the empty state is safe and the states below a safe state are safe. A multicast request is routed one by one, hence only one output is routed at a time. Let S be a safe state and let (v, w) be an (F, f)-limited request in S. Let $S(H)$ and $S(G)$ be the induced states.

Suppose v is idle in S. Since v is F-limited, there exists a route (v, u) in $S(H)$ where u is idle. The facts that u is idle and (v, w) is (F, f)-limited imply that (u, w) is (F, f)-limited. Hence there exists a safe state above $S(G)$ in G containing a route from u to w (see Fig. 4.2.2(a)).

Suppose v already has a route in S. Let (v, u) be the part of that route in $S(H)$. Then every route (u, z) in $S(G)$ corresponds to a route (v, z) in S. Since (v, w) is (F, f)-limited in S, (u, w) must be (F, f)-limited in $S(G)$. Hence there exists a safe state above $S(G)$ in G containing a route (u, w) (see Fig. 4.2.2(b)). $\square$

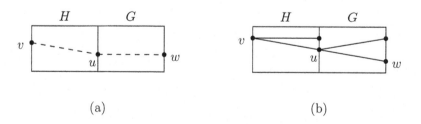

Figure 4.2.2: Concatenated paths

Theorem 4.2.9. *There exists a WSNB 3-stage network with closed-end multicast traffic and cost $O(N^{5/3})$.*

Proof. Set $N = 5^{3\lambda}$, $q = 5^{\lambda}$ and $F = 2 \cdot 5^{2\lambda-2}$. Let X be a $q^2 \times F$ crossbar. Then X is WSNB (F, f)-limited (q^2, F)-network for any f. By Lemmas 4.2.7 and 4.2.8, $G(q) \circ X$ is a WSNB (F, f)-limited (q^3, F)-network. The cost of $G(q) \circ X$ is easily verified to be $O(N^{4/3})$.

Set $t = \lceil N/F \rceil = O(N^{1/3})$. Vertically stacking t copies of $G(q) \circ X$ and identifying their inputs, we obtain a (N, tF)-multicast network with cost $O(N^{5/3})$. Delete all but N outputs. $\square$

With a similar construction, Feldman–Friedman–Pippenger obtained a 4-stage network with cost $O(N^{11/7})$ for closed-end multicast traffic. A slightly different construction yields a 5-stage network with cost $O(N^{3/2})$.

Feldman-Friedman-Pippenger also gave a construction of a WSNB multicast concentrator by controlling the number of idle neighbors of many set of inputs to be sufficient to meet its maximum demand. For a set X of vertices, let $E(X)$ denote the set of neighbors of X.

Lemma 4.2.10. *A bipartite graph G is a WSNB Ft-limited f-cast concentrator if for every set X of x vertices, $1 \leq x \leq 2F$, $|E(X)| \geq (f + t)x$.*

Proof. For a state S and a set X of inputs, define

$$
\begin{aligned}
A_S(X) &= \text{number of idle neighbors of } X \text{ in } S, \\
B_S(v) &= f\text{-number of busy neighbors connected to } v, \\
B_S(X) &= \sum_{v \in X} B_S(v), \\
C_S(X) &= A_S(X) - B_S(X).
\end{aligned}
$$

Define S to be a safe state if $C_S(X) \geq 0$ for every X with $|X| = x \leq 2F$. Blum–Karp–Papadimitriou–Vornberger–Yannakakis (1981) proved that to determine whether S is safe is co-NP-complete. It is easily verified that the empty state is safe. For any state S,

Claim (i). $0 \geq C_S(X) \Rightarrow x < F$.

Proof. $0 \geq C_S(X) = A_S(X) - B_S(X) > [(f + t)x - Ft] - fx = (x - F)t \Rightarrow x < F$. $\square$

Claim (ii). C is submodular, i.e., $C(X \cup Y) + C(X \cap Y) \leq C(X) + C(Y)$.

Proof. Since B is a sum over X, B is modular. It suffices to prove that A is submodular.

Any busy neighbor shared by X and Y but itself not a neighbor of $X \cap Y$ is counted once in $A(X \cup Y) + A(X \cap Y)$ but twice in $A(X) + A(Y)$. All other busy neighbors are counted the same number of times in both sums. $\qquad\square$

Claim (iii). If S is safe, then $C_S(X) = C_S(Y) = 0 \Rightarrow C_S(X \cup Y) = 0$.

Proof. By Claim (i), $x < F$ and $y < F$. Hence $|X \cup Y| < 2F$. Since S is safe, $C_S(X \cup Y) \geq 0$. By Claim (ii),

$$C_S(X \cup Y) \leq C_S(X) + C_S(Y) = 0.$$

Hence $C_S(X \cup Y) = 0$. $\qquad\square$

Back to the proof of Lemma 4.2.10. We first prove that if S is safe, then for any input v, there exists an idle $w \in E(v)$ such that $S \cup \{v, w\}$ is safe. Suppose not. Then for every idle $w \in E(v)$, there exists a set X_w such that $C_{S \cup \{v,w\}}(X_w) < 0$. If $v \in X_w$, then $w \in E(X_w)$. It follows that

$$A_{S \cup \{v,w\}}(X_w) \geq A_S(X_w) - 1,$$
$$B_{S \cup \{v,w\}}(X_w) \geq B_S(X_w) - 1.$$

Hence

$$C_{S \cup \{v,w\}}(X_w) \geq C_S(X_w) \geq 0,$$

contradicting the assumption $C_{S \cup \{v,w\}}(X_w) < 0$. If $v \notin X_w$, then

$$A_{S \cup \{v,w\}}(X_w) \geq A_S(X_w) - 1,$$
$$B_{S \cup \{v,w\}}(X_w) = B_S(X_w).$$

Hence

$$0 > C_{S \cup \{v,w\}}(X_w) \geq C_S(X_w) - 1.$$

It follows that $C_S(X_w) = 0$. Define $X = \bigcup_w X_w$. By Claim (iii), $C_S(X) = 0$. By Claim (i), $x < F$. Hence $|X \cup \{v\}| \leq F$ and $C(X \cup \{v\}) \geq 0$. On the other hand, $v \notin X_w$ for every w, hence $v \notin X$. But every neighbor w of v is in X_w, hence in X. It follows

$$A_S(X \cup \{v\}) = A_S(X),$$

$$B_S(X \cup \{v\}) \geq B_S(X) + 1 \quad \text{(since } v \text{ is an } (Ft, f)\text{-limited request)}.$$

Hence

$$C_S(X \cup \{v\}) \le C_S(X) - 1 = -1\,,$$

a contradiction. Therefore $S \cup \{v, w\}$ is safe.

Next we prove that $S \setminus \{v, w\}$ is safe. Consider any X. If $v \notin X$, then

$$A_{S \setminus \{v,w\}}(X) = A_S(X) + 1,$$
$$B_{S \setminus \{v,w\}}(X) = B_S(X) + 1.$$

If $v \notin X$, then

$$A_{S \setminus \{v,w\}}(X) \ge A_S(X),$$
$$B_{S \setminus \{v,w\}}(X) = B_S(X).$$

In either case $C_{S \setminus \{v,w\}}(X) \ge C_S(X) \ge 0.$ □

Theorem 4.2.11. *For every $s \ge 1$ and $N \to \infty$, a WSNB s-stage N-network with cost $O\left(N^{1+\frac{1}{s}}(\log N)^{1-\frac{1}{s}}\right)$ exists for closed-end multicast traffic.*

Proof. By a probabilistic argument, there exists a bipartite graph B with N inputs and $8F$ outputs such that every set of x inputs, $1 \le x \le 2F$, has at least $2x$ neighbors, and every output has $O(\log N)$ edges.

Theorem 4.2.11 is trivially true for $s = 1$. The general s case is proved by induction. Set $F = \left\lceil N^{1-\frac{1}{s}}(\log N)^{\frac{1}{s}} \right\rceil$ in the probabilistic argument, and set $f = t = 1$ in Lemma 4.2.10, we obtain a WSNB F-limited concentrator C with N inputs, $8F$ outputs, depth 1 and cost $O(N \log N)$. By induction, there exists a $8F$-multicast network with depth $s - 1$ and cost $O(F^{1+\frac{1}{s-1}}(\log F)^{1-\frac{1}{k-1}})$. By Lemma 4.2.8, $C \circ G$ is a WSNB $(N, 8F)$-multicast network with depth k and cost $O(N \log N)$. Let $t = \lceil N/8F \rceil = O\left(N^{\frac{1}{k}}/(\log N)^{\frac{1}{k}}\right)$. Vertically stacking t copies of $C \circ G$ and identifying their inputs, we obtain a WSNB $(N, 8Ft)$-multicast network with depth s and cost $O(N^{1+\frac{1}{k}}(\log N)^{1-\frac{1}{k}})$. Delete all but N outputs.   □

One can replace the bipartite graph B obtained by probabilistic argument by expanders with explicit construction. The cost of the corresponding network is $O(N^{1+\frac{2}{s}})$.

Unfortunately, neither the explicit nor the nonexplicit construction is practical since the problem of determining whether a state is safe in the routing algorithm is NP-complete. Nevertheless, Theorem 4.2.11 gives the connect best, costwise, s-stage multicast network.

Tscha and Lee (2001) proposed a multicast WSNB algorithm for the $\log_2(N, 0, m)$ network. Suppose the $N = 2^n$ outputs are labeled by binary n-sequences. Then a θ-window consists of the 2^θ outputs where labels may differ only in the last θ bits. An f-cast request will be split to several f-cast subrequests each consisting of outputs in a given θ-window. Two rules are observed in this θ-window routing :

(i) Each subrequest uses one path up to stage $s - \theta$(for an s-stage network).

(ii) The subrequests from the same request are treated as independent requests, i.e., they cannot share any link.

Tscha and Lee set $\theta = \lfloor n/2 \rfloor$ and proved.

Theorem 4.2.12. $\log_2(N, 0, m)$ *is multicast WSNB under the* $\lfloor n/2 \rfloor$-*window routing if* $m \geq \lfloor n/2 \rfloor 2^{\lfloor (n-1)/2 \rfloor} + 1$.

We hold the proof since a more general case will be proved. Tscha and Lee claimed their result to be SNB, but Kabacinski and Danilewitz (2002) pointed out the use of the window routing implies WSNB. They extended Theorem 4.2.12 to the case that θ is not fixed at $\lfloor n/2 \rfloor$, and proved

Theorem 4.2.13. $\log_2(N, 0, m)$ *is multicast WSNB under the* θ-*window routing if*

$$m \geq \begin{cases} \theta 2^{n-\theta-1} + \lceil 2^{n-2\theta-1} \rceil, & \text{for } 1 \leq \theta \leq \lfloor n/2 \rfloor, \\ 2^\theta + (n - \theta - 2)2^{n-\theta-1} - 2^{2\theta-n-1} + 1, & \text{for } \lfloor n/2 \rfloor + 1 \leq \theta \leq n. \end{cases}$$

Recently, Hwang and Lin (2003) further extended Theorem 4.2.13 to $\log_2 (N, k, m)$. The analysis for the general case is much more complicated as the connections between an input and an output is not unique. First of all, one has to be more specific about the window algorithm. Hwang and Lin proposed the delayed-splitting window algorithm which prescribes that a multicast connection to outputs in the same θ-window cannot be split before stage $(n+k-\theta+1)$. Note that further delay is not always possible since stage $n + k - \theta + 1$ is the last stage all outputs in the same window have common reachable crossbars. Also note that such an algorithm fixes only the relative routing of two outputs in the same θ'-window, $\theta' \leq \theta$, but not the absolute routing to an output. Thus whether two connections intersect is uncertain and the notion of intersecting graph used by Tscha and Lee is not applicable. Instead, Hwang and Lin adopted the method of channel graph blockage analysis first proposed by Shyy and Lea for single-cast.

Count all i-intersecting connections, $n + m - \theta \leq i \leq n + m - 1$, from the output side. Note that the outputs of these connections must all be in the designated θ-window. Thus there are at most $2^\theta - k$ of such connections. Further, they have different impacts in blocking the paths in the channel graph depending on i.

On the other hand, count all i-intersecting connections, $1 \leq i \leq n+k-\theta-1$, from the input side. Again, an i-intersecting connection has a greater (or equality permitted) blocking impact than an $(i + 1)$-intersecting calls for $i \leq \lfloor (n + k)/2 \rfloor$. It will be shown that there is no need to count from the input side beyond the stage $\lfloor (n + k)/2 \rfloor$. Therefore, the counting will be from small i to large i to maximize the blocking impact. We consider two cases:

(i) $0 \leq k \leq 1$

The number of stage-i links, $1 \leq i \leq n + k - 1$, in the channel graph is constant, 1 for $m = 0$ and 2 for $m = 1$. Therefore each intersecting connection has the same impact, regardless of which stage it intersects. The worst case occurs when there is a maximum number of intersecting connections, i.e., $2^\theta - 1$ from the designated window which cause a blocking of $(2^\theta - 1)/2^k$ copies.

(ii) $2 \leq k$

Let R denote the part of the new request going to a designated window and $|R| = f$. Suppose a 1-window contains r outputs in R. Then it can block at most

$2 \times \frac{1}{4} = \frac{1}{2}$ if $r = 0$ (only for the 1-window which is in the designated 2-window)

$1 \times \frac{1}{2} = \frac{1}{2}$ if $r = 1$,

$\qquad = 0$ if $r = 2$.

Therefore a 1-window can block at most $1/2$ copy of the channel graph. Consequently, a θ-window can block at most $2^{\theta-2}$ copies, which is achieved by having either $f = 2^{\theta-1}$ (each 1-window has $r = 1$) or $f = 2^{\theta-2}$ (half of the 1-windows have $r = 1$ and half have $r = 0$).

To count i-intersecting connections for $1 \leq i \leq n + k - \theta - 1$, consider two cases:

(i) $\theta \leq \lfloor (n+k)/2 \rfloor - 1$

The argument for this part is a straightforward extension of the argument by Kabacinski and Danilewicz (2002) for $k = 0$. There are 2^{i-1} inputs which can generate an i-intersecting connection. Further, an i-intersecting connection can reach all windows for $i \leq k$, and $2^{n-\theta-i+k}$ windows for $i \geq k$. In the worst-case scenario, an i-intersecting connection is a multicast connection going to one output in each window it can reach except the designated window for $1 \leq i \leq \theta$. The reason for exception is that all outputs in the designated window are already counted in the part concerning $n + k - \theta \leq i \leq n + k - 1$. Since an i-intersecting connection blocks 2^{-i} copies for $i \leq k$ and 2^{-k} copies for $k \leq i \leq \lfloor (n+k)/2 \rfloor$, the total blockage of up to stage θ is

$$\sum_{i=1}^{\theta} 2^{i-1}(2^{n-\theta} - 1)2^{-i}$$

$$= \sum_{i=1}^{\theta} 2^{n-\theta-1} - \sum_{i=1}^{\theta} 2^{-1}$$

$$= \theta(2^{n-\theta-1} - 1/2) \qquad \text{for } \theta \leq k$$

and

$$\sum_{i=1}^{k} 2^{i-1}(2^{n-\theta} - 1)2^{-i} + \sum_{i=k+1}^{\theta} 2^{i-1}(2^{n-\theta-i+k} - 1)2^{-k}$$

$$= \sum_{i=1}^{k} 2^{n-\theta-1} - \sum_{i=1}^{m} 2^{-1} + \sum_{i=k+1}^{\theta} 2^{n-\theta-1} - \sum_{i=k+1}^{\theta} 2^{i-k-1}$$

$$= \theta 2^{n-\theta-1} - k/2 - 2^{\theta-k} + 1 \qquad \text{for } \theta \geq k.$$

Note that these i-intersecting connections, $1 \leq i \leq \theta$, use up a maximum of $\sum_{i=1}^{\theta} 2^{i-1} = 2^{\theta} - 1$ outputs in a window. Therefore one $(\theta + 1)$-intersecting connection can still fit in if $\theta + 1 < n + k - \theta$, or $\theta \leq \lfloor (n+k)/2 \rfloor - 1$, which is the case here. This $(\theta + 1)$-intersecting connection reaches $2^{n-\theta} - 1$ windows for $\theta < m$, and $2^{n-2\theta-1+k} - 1$ windows for $\theta \geq m$, while each path to a window blocks 2^{-k} copy.

To summarize, the number of blockages from the input side is

$$\theta(2^{n-\theta-1} - \frac{1}{2}) + 2^{n-\theta-k} - 2^{-k} \qquad \text{for } \theta < k$$

$$\theta 2^{n-\theta-1} - k/2 - 2^{\theta-k} + 1 + 2^{n-2\theta-1} - 2^{-k} \qquad \text{for } \theta \geq k.$$

(ii) $\theta \geq \lfloor (n+k)/2 \rfloor$

Then $\theta \geq k$. Note that i-intersecting connections for $n + k - \theta \leq i \leq n + k - 1$ are counted from the output side. So the input side counts only up to stage $n + k - \theta - 1$ (which is upper bounded by θ). Thus the number of blockings from the input side is

$$\sum_{i=1}^{k} 2^{i-1}(2^{n-\theta} - 1)2^{-i} + \sum_{i=k+1}^{n+k-\theta-1} 2^{i-1}(2^{n-\theta-i+k} - 1)2^{-k}$$
$$= (n + k - \theta - 1)2^{n-\theta-1} - k/2 - 2^{n-\theta-1} + 1$$
$$= (n + k - \theta - 2)2^{n-\theta-1} - k/2 + 1.$$

Since each intersecting connection counted from the output side blocks in the worst-case scenario, i.e., $f = 2^{\theta-1}$ or $2^{\theta-2}$, at least $1/4$ copy, there is no reason for the counting from input side to go over stage $n + m - \theta$ with one exception.

For $\theta \geq 2$, we can increase the blocking by allowing the unique 1-intersecting connection from the input side to also go to the designated window to reach an output blocking $1/4$ copy (such an output exists when $k = 2^{\theta-2}$). Then this intersecting connection blocks $1/2$ copy if counted from the input side, greater than its original value $1/4$ as counted from the output side. Note that no other such reversal of counting will bring any further increase since the 1-intersecting connection is the only one which blocks more than $1/4$ copy when counted from the input side. On the other hand, since all intersecting connections counted from the input side are before the middle stage, reverse them to the output side will only decrease their impact on depletion.

Combining the above, we have

Theorem 4.2.14. *$Log_2(N, m, p)$ is WSNB for broadcast under the θ-window* *algorithm if and only if*

$$m > \begin{cases} \theta \cdot 2^{n-\theta-1} + 2^{n-2\theta-1} - 1, & for \ k = 0, \theta \leq \lfloor n/2 \rfloor - 1, \\[4pt] \theta \cdot 2^{n-\theta-1} + 2^{n-2\theta-1} - 1/2, & for \ k = 1, \theta \leq \lfloor (n+1)/2 \rfloor - 1, \\[4pt] 2^{\theta} + (n - \theta - 2) \cdot 2^{n-\theta-1}, & for \ k = 0, \lfloor n/2 \rfloor \leq \theta \leq n - 1, \\[4pt] 2^{\theta-1} + (n - \theta - 1) \cdot 2^{n-\theta-1}, & for \ k = 1, \lfloor (n+1)/2 \rfloor \leq \theta \leq n - 1, \\[4pt] 2^{\theta-2} + \theta \cdot 2^{n-\theta-1} - k/2 - 2^{\theta-k} + 2^{n-2\theta-1} - 2^{-k} + 5/4, \\ \qquad for \ 2 \leq k \leq \theta \leq \lfloor (n+k)/2 \rfloor - 1, \\[4pt] 2^{\theta-2} + \theta \cdot 2^{n-\theta-1} - \theta/2 + 2^{n-\theta-k} - 2^{-k} + 1/4(0 \ if \ \theta = 1), \\ \qquad for \ k > \max\{\theta, 1\}, \theta \leq \lfloor (n+k)/2 \rfloor - 1, \\[4pt] 2^{\theta-2} + (n + k - \theta - 2) \cdot 2^{n-\theta-1} - k/2 + 5/4, \\ \qquad for \ 2 \leq k \leq \lfloor (n+k)/2 \rfloor \leq \theta \leq n - 1. \end{cases}$$

Note that $Log_2(N, n - 1, p)$ is the Cantor network.

Corollary 4.2.15. *The Cantor network is WSNB for broadcast under the window algorithm if and only if* $p > 2^{\theta-2} + \theta \cdot 2^{n-\theta-1} - \theta/2 + 2^{1-\theta} - 2^{1-n} + 1/4(0 \ if \ \theta = 1)$.

Hwang and Lin also showed that to minimize cost, θ does not have to exceed 2, and $\theta = 2$ is generally the best. They gave a method to optimize m for $\theta = 0, 1, 2$.

Hwang (2003) proposed the θ-neighborhood algorithm to 3-stage Clos network. Partition the output switches into sets of t (assuming t divides r_2), call each set a *window*. An k-request involving outputs in w windows will be treated as w multicast requests, where the ith sub-request involves only the k_i outputs in window i. Each sub-request must be routed through one middle switch only, while two sub-requests from the same input are treated as different requests and hence cannot be routed through the same middle switch.

Note that the window algorithm implies the no-split rule since two outputs on the same output switch must be in the same window and their routes go through the same middle switch. Clearly, their routes must also go through the same link between that middle switch and their output switch. Therefore, in deriving the tightest nonblocking condition, we may assume that each k-request involves at most one output from an output switch. Hence k_i is also the number of output switches in window i involved in the k-request. Clearly, $w \leq \min\{k, r_2/t\}$, $k_i \leq \min\{t, k - w + 1\}$ and $k \leq \min\{f, r_2\}$.

Theorem 4.2.16. *For the t-window algorithm under the closed-end traffic:*

$$m^o = \max\{\min\{n_1 \min\{f, r_2/t\} + (\lfloor v^o \rfloor + 1)(n_2 - 1), N_1 \min\{f, r_2/t\}\},$$
$$\min\{N_2 - \lceil v^o \rceil, N_1 \min\{f, r_2/t\}\}\},$$

where

$$v^o = \min\{[N_2 - (n_1 - 1)\min\{f, r_2/t\} - (n_2 - 1) - w]/n_2, t - 1, f - w\}.$$

Further, if $\lfloor v^o \rfloor = \lceil v^o \rceil$ *then*

$$m^o = \min\{(n_1 - 1)\min\{f, r_2/t\} + (v^o + 1)(n_2 - 1) + w,$$
$$(N_1 - 1)\min\{f, r_2/t\} + w, N_2 - v^o\}.$$

Proof. Suppose the request c consists of w sub-requests. Note that each sub-request involves a distinct set of outputs. We compute the maximum number of middle switches needed to connect sub-request i. Since each other sub-request is connected through a single middle switch, the connections of the other $w - 1$ sub-requests consume $w - 1$ middle switches regardless of the numbers of outputs in these sub-requests as long as they are positive. Therefore the worst case for sub-request i occurs when k_i is maximum, with k and w fixed, i.e., $k_i = \min\{t, k - w + 1\}$. Since the sub-requests are interchangeable, the worst-case number of middle switches suffices for one sub-request suffices for all.

Since each sub-request is routed through a single middle switch, Eq. (4.1.5) for model 1 can be used to compute the number of middle switches required to connect sub-request i, except replacing k by k_i. To compute m^o for connecting c, we also have to maximize over k and w, and add the $w - 1$ middle switches required to connect the other $w - 1$ sub-requests. Since c consists of at most $\min\{f, r_2/t\}$ sub-requests, we have $b(I) = (n_1 - 1)\min\{f, r_2/t\}$, $b(N_1) = (N_1 - n_1)\min\{f, r_2/t\}$, $b(c) = 1$ and $w \leq \min\{f, \lceil r_2/t \rceil\}$.

$$
\begin{aligned}
m^o &= \max_{k,w,k_i} [\min\{(n_1 - 1)\min\{f, r_2/t\} + k_i(n_2 - 1) + 1, \\
&\quad (N_1 - 1)\min\{f, r_2/t\} + 1, N_2 - k + 1\}] + w - 1 \\
&= \max_{k,w} [\min\{(n_1 - 1)\min\{f, r_2/t\} + \min\{t, k - w + 1\}(n_2 - 1) + w, \\
&\quad (N_1 - 1)\min\{f, r_2/t\} + w, N_2 - k + w\}] \\
&= \max_{\substack{0 \le v \le t-1 \\ 1 \le w \le \min\{f, r_2/t\}}} [\min\{(n_1 - 1)\min\{f, r_2/t\} + (v + 1)(n_2 - 1) + w, \\
&\quad (N_1 - 1)\min\{f, r_2/t\} + w, N_2 - v\}]
\end{aligned}
$$

$$(4.2.1)$$

by changing variable from $k - w$ to v. Setting the two terms in (4.2.1) containing v equal, we obtain the first term of v^o. The second and third terms represent some boundary conditions imposed by the other two terms in (4.2.1).

Clearly, the maximum is obtained at $v =$ either $\lfloor v^o \rfloor$ or $\lceil v^o \rceil$, and at $w = \max w = \min\{f, r_2/t\}$. At $v = \lfloor v^o \rfloor$, we can drop the third term of (4.2.1) as it is larger than the first term. At $v = \lceil v^o \rceil$, we can drop the first term.

Thus Theorem 4.2.16 follows. □

Corollary 4.2.17. *For $t = r_2$, $m^o = \min\{N_1, N_2 - (N_2 - n_1 + 1)/n_2 + 1\}$.*

Note that m^o in Corollary 4.2.17 is same as in model 1. This is not surprising. Since if all output switches are in the same window, then every f-cast request must be routed through the same middle switch, the same constraint as in model 1.

Corollary 4.2.18. *For $t = 1$, $m^o = \min\{r_1 \min\{f, r_2\} + n_2 - 1, N_1 \min\{f, r_2\}, N_2\}$.*

Suppose $r_2/t \le t$. Then the maximums of intervals (a) and (c) are also in interval (b). Hence the maximum of interval (b) is the overall maximum.

Suppose $t \le r_2/t$. Then interval (b) does not exist while the maximum of interval (c) is also in interval (a). Hence the maximum of interval (a) is the overall maximum. □

The m^o in Corollary 4.2.18 is same as in model 2 except that f is replaced by $\min\{f, r_2\}$. This difference is due to the fact that no-split is forced under the window algorithm but not in model 2.

Figure 4.1.2 can also be used to show that $C(n_1, r_1, n_2, r_2, m)$ cannot guarantee WSNB for open-end traffic under the window algorithm no matter how large m is. Suppose the two output switches with outputs o_1, o_2, o_3, o_4 are in the same window (other output switches can be added). Then request (i_1, o_2) is unroutable since it must be routed through the first middle switch under the window algorithm, but the link from that middle switch to the first output switch is occupied.

The window algorithm offers a continuum of choices between model 1 and model 2. It also compares favorably with some other models in this section. As we said, model 1 corresponds to choosing $t = r_2$, and model 2 is close to but dominated by the choice $t = 1$. For f large (for example, the uncontrained broadcast case where $f = r_2$), models 0,2,3 and the no-split algorithm essentially require $n_1 r_2$ middle switches while the window algorithm requires $n_1 \sqrt{r_2}$ by choosing $t = \sqrt{r_2}$. On the other hand, for f_2 small, we can choose $t = 1$ such that the window algorithm essentially requires the same number of middle switches as models 0,2,3 and the no-split algorithm.

4.3   Rearrangeable Multicast Networks

Masson and Jordan (1972) first started the study of multicast networks in English literature. They proved (stated in the strengthened version of Hwang (1972)):

Theorem 4.3.1. $C(n_1, r_1, m, n_2, r_2)$ *under model 2 is multicast rearrangeable if and only if* $m \geq m^o$, *where*

$$m^o = \max\{\min\{n_1 f, N_2\}, \min\{n_2, N_1\}\}.$$

Proof. An f'-request can be treated as f' 1-requests since it has to be routed through f' distinct middle crossbars. With this understanding, then the maximum degree of an input node in the frame graph is $\Delta_1 = \min\{n_1 f, N_2\}$, and the maximum degree of an output node is $\Delta_2 = \min\{n_2, N_1\}$. Theorem 4.3.1 follows from Theorem 3.1.2.   □

Note that for rearrangeability, there is no distinction between closed-end and open-end traffic. Let $\varphi(n_1, r_1, m, n_2, r_2)$ be the minimum number of rearrangements needed for $C(n_1, r_1, m, n_2, r_2)$.

Corollary 4.3.2. $\varphi(n_1, r_1, m^o, n_2, r_2) = \min\{r_1, r_2\} - 1$.

By using the no-split algorithm, we can set $f = r_2 = n_1 = O(N^{1/3})$ in Theorem 4.3.1 and obtain a cost of $O(N^{5/3})$, which is the same as SNB or WSNB under model 2. Kirkpatrick–Klawe–Pippenger (1985) considered the case that the input stage has no fan-out capacity.

Theorem 4.3.3. *Consider model 1.*

(i) $C(n_1, r_1, m, n_2, r_2)$ *is RNB if* $m \geq [N_2(n_2 - 1)]^{1/2} + n_1 - n_2 + 1$.

(ii) $C(n_1, r_1, m, n_2, r_2)$ *is not RNB if* $m < \max\left\{ n_1, \dfrac{[N_2(n_2-1)]^{1/2}}{2} \right\}$.

Proof. (i). There are at most $[N_2(n_2 - 1)]^{1/2}$ requests each engaging at least $[N_2/(n_2 - 1)]^{1/2}$ outputs. These requests are routed first. Assign each such request to a distinct middle crossbar. Therefore $[N_2(n_2 - 1)]^{1/2}$ middle crossbars suffice to route these large requests. For any other request, the number of middle crossbars engaged by co-inputs and co-outputs are at most

$$n_1 - 1 + \left([N_2/(n_2 - 1)]^{1/2} - 1 \right)(n_2 - 1) \leq n_1 + [N_2(n_2 - 1)]^{1/2} - n_2.$$

So one more middle crossbar connects the request.

(ii). As $m < n_1$ is trivial, we prove $m < [N_2(n_2 - 1)]^{1/2}/2$. Represent a request by its set of output crossbars. It suffices to construct $[N_2(n_2-1)]^{1/2}/2$ requests with each pair intersecting. Let K be a complete graph with $\left\lceil r_2^{1/2} \right\rceil$ vertices. Label the vertices by request, and edges by output crossbars. Then the edges of a vertex are the requested output crossbars. Clearly, K yields a set of $\left\lceil r_2^{1/2} \right\rceil$ intersecting requests. $n_2/2$ copies of K will do as

$$\frac{n_2}{2} \left\lceil r_2^{1/2} \right\rceil \geq \frac{[n_2(n_2 - 1)]^{1/2}}{2} r_2^{1/2} = \frac{[N_2(n_2 - 1)]^{1/2}}{2}.$$

$\square$

By setting $n_1 = r_2 = O(N^{2/3})$ and $n_2 = r_1 = O(N^{1/3})$, the cost of (i) is $O(N^{5/3})$, better than the $O(N^2)$ cost of SNB or WSNB under model 1. Kirkpatrick–Klawe–Pippenger also considered the case where there is no constraint on the fan-out capacity. First, a lemma is needed.

Lemma 4.3.4. (i). $C(n_1, r_1, m, n_2, r_2)$ *is multicast RNB if* $m \geq n_1 n_2$, (ii). $C(n_1, r_1, m, n_2, r_2)$ *is not multicast rearrangeable if* $m \leq (n_1 - 1)n_2$, $r_1 \geq n_2\binom{(n_1-1)n_2}{n_2-1}$ *and* $r_2 \geq \binom{N_1}{n_2}$.

Proof. (i). Consider all requests from an input crossbar X. Partition the $n_1 n_2$ middle crossbars into n_1 disjoint sets of n_2 each. Since any set of n_2 middle crossbars not engaged by any co-input can route any request to any output crossbar, each such set can route a request from X. Since X is arbitrary, and the above argument does not depend on the routing of requests from other input crossbars, we are done. Note that the reservation of n_2 middle switches for each input on an input crossbar can be used as a WSNB algorithm.

(ii). Set $r_1 = n_2\binom{(n_1-1)n_2}{n_2-1}$ and $r_2 = \binom{N_1}{n_2}$ where each output crossbar receives n_2 requests from a distinct set of n_2 inputs. Suppose $m = (n_1 - 1)n_2$ and every input generates a request. Since a middle crossbar can carry connections from at most r_1 requests, at most $mr_1/n_2 = (n_1-1)r_1$ requests can each be routed through n_2 or more middle crossbars. In other words, at least $N_1 - (n_1 - 1)r_1 = r_1$ requests are each routed through $n_2 - 1$ or fewer middle crossbars. There are only $\binom{m}{n_2-1}$ distinct sets of $n_2 - 1$ middle crossbars. We say a request is covered by such a set $\mathcal{Y}$ if the middle crossbars used to connect that request all belong to $\mathcal{Y}$. Note that a request using fewer than $n_2 - 1$ middle crossbars can be covered by more than one set. Since $r_1/\binom{m}{n_2-1} = n_2$, there exists a set $\mathcal{Y}$ which covers at least n_2 requests. Choose any n_2 requests and call it $\mathcal{X}$. By construction, there exists an output crossbar $\mathcal{Z}$ receiving $\mathcal{X}$. Then there is no way to connect the n_2 outputs of $\mathcal{Z}$ to the $n_2 - 1$ middle crossbars in $\mathcal{Y}$ available to them.   □

Unlike results (i) and (ii) in Theorem 4.3.3, or result (i) in Lemma 4.3.4, result (ii) in Lemma 4.3.4 holds only for particular choices of r_1 and r_2. In other words, if $m \leq (n_1 - 1)n_2$, then there exist large r_1 and r_2 to prevent rearrangeability. Therefore $(n_1 - 1)n_2$ should not be interpreted as a lower bound of m to guarantee rearrangeability.

Theorem 4.3.5. $C(n_1, r_1, m, n_2, r_2)$ is multicast RNB if $m \geq (n_1-1)\lceil\log_2 r_2\rceil + 2n_2 - 1$.

Proof. (i) Suppose $\lceil\log_2 r_2\rceil \geq 2n_2 - 1$. Then $m \geq n_1(2n_2 - 1) \geq n_1 n_2$.

(ii) Suppose $\lceil\log_2 r_2\rceil \leq 2n_2 - 2$. Route the requests one by one by assigning at most $\lceil\log_2 r_2\rceil$ middle crossbars to connect one request routed. We prove by induction on the number of requests routed.

There is certainly no problem to route the first request. Consider the routing of the ith request, say from X to $Z_1, \ldots, Z_t$. Then there is at least a set M of $2n_2 - 1$ middle crossbars not engaged by the co-inputs of the new request. Choose any set $K \subseteq M$ of k middle crossbars. Then it cannot connect the requested outputs in Z_j only if these k crossbars are all engaged by the co-outputs of Z_j. But there are only $\binom{n_2-1}{k}$ such choices of K. Consider all $Z_1, \ldots, Z_t$, then there are at most $t\binom{n_2-1}{k} \leq r_2\binom{n_2-1}{k}$ choices of k middle crossbars which cannot connect the requested outputs from at least one of the Z_j. So if $\binom{2n_2-2}{k} > r_2\binom{n_2-1}{k}$, we are done. But for $k = \lceil\log_2 r_2\rceil \leq 2n_2 - 2$, we have $r_2\binom{n_2-1}{k} \leq 2^k\binom{n_2-1}{k} \leq \binom{2n_2-2}{k}$.   □

By setting $n_1 = r_2 = (N/\log N)^{1/2}$ and $n_2 = r_1 = (N \log N)^{1/2}$, the cost of this network is $O(N^{3/2}(\log N)^{1/2})$, which is slightly worse than that of the Yang–Masson 3-stage WSNB multicast network. The proof actually defines a WSNB algorithm for open-end traffic. Case (ii) is same as the Yang–Masson WSNB algorithm by setting $p = \lceil \log_2 r_2 \rceil$ (but the argument is different). Therefore, Yang–Masson's algorithm of finding a set of middle crossbars which can accommodate a request also applies to Case (ii). The routing of Case (i) is trivial by routing all requests from the ith input of any input switch to the jth output of any output switch through the middle crossbars labeled (i, j), $i = 1, \ldots, n_1$, and $j = 1, \ldots, n_2$.

Richards–Hwang (1985) considered a variation of the 3-stage Clos network, which they called a *2-stage network*. Suppose an input crossbar X is of size $n \times m$ where n divides m, say $m/n = M$. If the m outlets of X are partitioned into n groups of M each, and each group is dedicated to the connection of a particular input of X, then essentially, X is replaced by n $1 \times M$ crossbars. Since an $1 \times M$ crossbar can be eliminated if the M outlets (which become the M inlets of the next stage) are each labeled by the input, the first stage can be eliminated and each network input is assigned to M middle crossbars (one inlet each), which become the input switches of the new 2-stage network. What is lost is the switching capability of the first stage of the 3-stage network. What is gained is a balanced allocation of resources (the middle-stage inlets) to the network input. Recall that blocking occurs in a 3-stage network often because a certain input has no access to any middle-stage inlet.

It turns out that not only a balanced number of occurrences of each input is good for RNB, but also a balanced number of joint occurrences of two inputs on the same input crossbar is. Treating the inputs as elements, and the input switches as blocks, then the balancedness condition is exactly what a block design prescribes. However, the network parameters typically do not fit the design parameters. So "balancedness" is compromised to mean "as balanced as possible," i.e., two joint occurrences can differ in number by at most 1. For most practical networks, the joint occurrence number is either zero or one.

Richards–Hwang suggested the following assignment scheme L. Suppose $\sqrt{N_1}$ is a prime. Then there exist $\sqrt{N_1} - 1$ orthogonal latin squares $S_1, \ldots,$ $S_{\sqrt{N_1}-1}$, of order $\sqrt{N_1}$. Let S_0 denote the $\sqrt{N_1} \times \sqrt{N_1}$ square where every entry of row i is i, and let $S_{\sqrt{N_1}}$ denote the transpose of S_0. Define $\bar{S}_i$ to be the subarray obtained by superimposing S_i on S_0, $i = 1, \ldots, \sqrt{N_1}$. Then each cell of $\bar{S}_i$ contains an ordered pair, which, when interpreted as a number of base $\sqrt{N_1}$, yields a distinct number. Define $\bar{S}_{\sqrt{N_1}+1}$ to be the transpose of

$\bar{S}_{\sqrt{N_1}}$. Then for a given $M \leq \sqrt{N_1} + 1$, we can choose any M of the $\sqrt{N_1} + 1$ subarrays, treating the entries as inputs and rows as input crossbars, to obtain the input assignment for a 2-stage network in which every two inputs occur on the same input crossbar at most once. Figure 4.3.1 gives the four subarrays for $N_1 = 9$.

0 1 2	0 1 2	0 2 1	0 0 0	0 4 8	0 7 5	0 1 2	0 3 6
0 1 2	1 2 0	1 0 2	1 1 1	3 7 2	3 1 8	3 4 5	1 4 7
0 1 2	2 0 1	2 1 0	2 2 2	6 1 5	6 4 2	6 7 8	2 5 8
S_0	S_1	S_2	S_3	$\bar{S}_1$	$\bar{S}_2$	$\bar{S}_3$	$\bar{S}_4$

Figure 4.3.1: The super-squares of $N_1 = 9$

Let $C_A(M, n_1, r_1, n_2, r_2)$ denote the 2-stage network under the assignment A. Then $N_1 = n_1 r_1 / M$ and $N_2 = n_2 r_2$. Then $C(M, n_1, r_1, n_2, r_2)$ denotes the network structure before any assignment. Since outputs of different output crossbars do not compete for links, the rearrangeability of $C(M, n_1, r_1, n_2, r_2)$ is independent of r_2. Let n_2^A denote the n_2 such that $C_A(M, n_1, r_1, n_2, r_2)$ is multicast rearrangeable under the assignment scheme A if and only if $n_2 \leq n_2^A$. In general, we can only obtain bounds for n_2^A. To prove that b is a lower bound of n_2^A, we need to show that for any set of b inputs, there exist b input crossbars such that every input is contained in a distinct input crossbar. By Hall's theorem (1935) on distinct representatives, it suffices to prove that for any set K of k, $k \leq b$, inputs, $|R| \geq k$ where R is the set of rows intersecting K (i.e., containing an input of K).

An assignment also defines a bipartite graph G with the inputs and the input crossbars as vertices, and assignments of inputs to input crossbars as edges. Then the condition on b being a lower bound of n_2^A can be translated to that G is a (N_1, r_1, b)-partial concentrator.

Richards–Hwang proved

Theorem 4.3.6. $n_2^L \geq M^2 + M - 1$.

Proof. For a given set K of $k \leq M^2 + M - 1$ inputs, let R_i denote the set of rows in subarray i which intersect K. It suffices to prove $\sum_{i=1}^{M} |R_i| \geq k$. Without loss of generality, assume $|R_1| \leq |R_2| \leq \cdots \leq |R_M|$. Let $|R_1| = |R_2| - d$. Since every two inputs occur in a row at most once, it follows that every two rows can share at most one input. Hence $k \leq |R_1| \cdot |R_2|$. Furthermore,

$$\sum_{i=1}^{M} |R_i| \geq |R_2| - d + (M-1)|R_2| = M|R_2| - d.$$

To prove $\sum_{i=1}^{M} |R_i| \geq k$, we consider two cases:

(i) $d \geq \frac{|R_2|(|R_2|-M)}{|R_2|-1}$.

$$M|R_2| - d \geq (|R_2| - d)|R_2| \geq k.$$

(ii) $d < \frac{|R_2|(|R_2|-M)}{|R_2|-1}$.

Since $d \geq 0$ implies $|R_2| > M$, we have $|R_2| \geq M + 1$.

$$M|R_2| - d \geq M|R_2| - \frac{|R_2|(|R_2| - M)}{|R_2| - 1} = \frac{|R_2|^2(M-1)}{|R_2| - 1},$$

which increases in $|R_2|$ for $|R_2| > 2$. Therefore

$$M|R_2| - d > \frac{(M+1)^2(M-1)}{M} = M^2 + M - 1 - \frac{1}{M},$$

or

$$M|R_2| - d \geq M^2 + M - 1 \geq k.$$

$\square$

Hwang–Richards (1992) later improved Theorem 4.3.6 to $n_2^L \geq M^2 + 2M - 2$. On the other hand, Du–Hwang–Richards (1985) showed that there exists a choice H of M subarrays in the scheme L such that $n_2^{L(H)} \leq (2/3)M^3 - 2M^2 + (16/3)M - 3$. Later, Hu–Hwang (1992) found another choice H' such that $n_2^{L(H')} \leq (2/3)M^3 - 2M^2 + (10/3)M + 5$. It was suggested that H and H' are both close to the worst choice of M subarrays, while a good choice, yet unknown, will lead to a much higher upper bound.

The cost of $C(M, n_1, r_1, n_2, r_2)$ is

$$r_1 n_1 r_2 + r_2 n_2 r_1 = N_1 N_2 \left(\frac{M}{n_2} + \frac{r_1}{N_1} \right).$$

M/n_2 is called the *capacity ratio*, r_1/N_1 the *concentration ratio*, and their sum the *cost ratio*. These ratios are determined solely by the partial concentrator.

Suppose $n_2 = O(M^2)$ as the lower bound of n_2^L suggests. Then by setting $n_2 = \sqrt{N_1} = \sqrt{N}$, the cost is $O(N^{7/4})$. Suppose $n_2 = O(M^3)$ as the upper bound suggests. Then by setting $n_2 = \sqrt{N_1} = \sqrt{N}$, the cost is $O(N^{5/3})$.

If $\sqrt{N_1}$ is not a prime, let f be its smallest factor. Then there exist $f + 1$ subarrays. If N_1 is not a square, then one has to go to the next square number.

For dynamic traffic routing is trivial for the 2-stage network. A table giving the set I_i of input crossbars containing input i can be stored in each output switch. A request is broken down to its point-to-point component (i, j) and considered being generated by the output j, which simply checks its table and tries the members of I_i in some sequential order. Since the subarrays in the scheme L are constructed algebraically, one can also install a formula, instead of the table, and compute the set I_i.

For scheduled traffic, the requests of each output switch can be routed independently. The routing of an output switch is a system of distinct representatives problem with n_2 M-subsets.

Suppose the request (i, j) is blocked. A rearrangement tree is constructed where the root (level 0) is labeled by i. i has M sons labeled by the inputs each of which has a connection to a co-output carried by a member of I_i. Each of these sons k has $M - 1$ sons of their own either unlabeled or labeled by inputs which have connections to co-outputs carried by a member of I_k. If k has an unlabeled son, then k's connection can be moved to that available input crossbar, and i takes k's place. If all grandsons of i are labeled, we have to move onto the next level until an unlabeled descendant shows up. Note that level $\ell \geq 2$ has $M(M-1)^{\ell-1}$ descendants. On the other hand, the labels are all inputs requested by the $n_2 - 1$ co-outputs of j. In theory, a label can appear $n_1 - 1$ times in the tree, but the balancedness properties of the assignment scheme usually prevent it from appearing often. So the number of descendants at a level catches up fast.

Jacobsen (1988) formalized this by defining $S(n) = $ minimum number of input crossbars intersecting a set of n inputs, $S(0) = 0$ and $S(k+1) = S(S(k)+1)$. Then $S(k)$ is a lower bound on the number of distinct inputs labeling the first k levels. Thus an upper bound on the number of rearrangements is the smallest k such that $S(k) \geq n_2$. For $M = 2$, this number is 2.

Pinsker (1973) and Pippenger (1973) first introduced the notion of partial concentrator and gave some probabilistic constructions. Since then, various combinatorial objects have been used to devise partial concentrators. Here we list a few:

Binomial (B) by Masson (1977).

Orthogonal array (OA) by Kufta–Vacroux (1983) and Richards–Hwang (1985).

Subarray (S) by Richards–Hwang (1985) and Hwang–Richards (1992).

Hypercube (H) by Du–Hwang–Richards (1985) and West–Benerjee (n.d.).

Multiple binomial (MB) by West–Benerjee.

2-design (T) by Hwang–Richards (1991).

Resolvable-t-design (R) by Hwang–Richards (1991).

Double-intersect (D) by Hwang–Richards (1991).

Figure 4.3.2 gives the parameter and properties of these partial concentrators.

Type	N_1	r_1	n_2	conditions	Cap. Ratio	Con. Ratio
B	$\frac{r_1}{M}$	r_1	$M+2$		$M/(M+2)$	$(N_1 M^M)^{\frac{1}{M}-1}$
OA	s^2	Ms	M^2+M-1	$OA(s,M)$ exists	$M/(M^2+M-1)$	M/s
S	s^2	Ms	M^2+2M-2	s a prime	$M/(M^2+2M-2)$	M/s
H	s^d	$s^{d-1}M$	M^d	$d\|M$	$1/M^{d-1}$	M/s
MB	$M\frac{r_1}{M}$	r_1	$M+1$		$M/(M+1)$	$(N_1 M^M)^{\frac{1}{M}-1}$
T	N_1	r_1	M^2/λ	a $(r_1, N_1, MN_1/r_1, M_1, \lambda)$ 2-design	λ/M	r_1/N_1
R	N_1	r_1	M^2/λ	a $(N_1, r_1, M, Mr_1/N_1, 1)$ resolvable t-design	$t/M+1$	$M^{t/(t-1)}/N_1$
D	N_1	r_1	$(M^3-M^2)/\lambda+M$	a $\frac{r_1}{M}, \frac{N_1 M}{r_1}, \lambda$ difference set	$\lambda/(M^2-M+\lambda)$	$M/(\lambda N_1)^{1/3}$

Figure 4.3.2: A list of partial concentrators

For fixed M, the capacity ratio usually dominates the concentration ratio and determines the cost complexity. Note that the capacity ratios in Fig. 4.3.2 are mostly constant or M^{-1}, except D is of the order M^{-2} and H is M^{-d+1} (d must divide M). Therefore, the capacity ratio is always bounded away from zero. For $M = 2$, Garey–Hwang–Richards (1988) related capacity to the girth of the corresponding bipartite graph and showed that the cost ratio $= O(\log \log N_1 / \log N_1)$. For $M \geq 3$, they gave a probabilistic argument that a partial concentrator with cost ratio $O(N_1^{-(M-2)/2(M-1)})$ exists.

A recursive extension of both the Masson–Jordan 3-stage network and the Richards–Hwang 2-stage network to multistage yields an $O(N^{3/2})$-cost network. A recursive extension of the Kirkpatrick–Klawe–Pippenger network yields a $(2k+1)$-stage multicast RNB network with cost $O(N^{1+\frac{1}{k+1}}(\log N)^{\frac{k}{2}})$ network. Dolev–Dwork–Pippenger–Widgerson also gave a nonconstructive multicast rearrangeable s-stage network with cost $O\left((N\log N)^{1+\frac{1}{k}}\right)$.

Ofman (1965) proposed the approach to construct a multicast network by concatenating two networks each performing one of the two functions of a multicast network. Namely, a multicast concentrator (also called a *copy network*) allows the input of an f-request to reach some f outputs, and a point-to-point network to connect the f inputs (which are outputs of the first network) to their proper destinations.

Ofman showed that a copy network could be constructed by concatenating a hyperconcentrator with a multicast infra-concentrator. He then proved that BY^{-1} is a multicast infra-concentrator. Here we state the result for any infra-connector which BY^{-1} is.

Theorem 4.3.7. *An infra-connector is a multicast infra-concentrator.*

Proof. Let ν denote an infra-connector. Let p_i denote the number of copies requested by i, $1 \le i \le k$, in π. Set $P_i = \sum_{j=1}^{i} p_j$. Then ν can route the infra-frame $\bar{\pi} = \{(i, P_i) : 1 \le i \le k\}$. Furthermore, if for each i we change P_i to $P_i' = P_i - y_i$ for some $0 \le y_i \le p_i - 1$. Then $\bar{\pi}' = \{(i, P_i') : 1 \le i \le k\}$ is still an infra-frame and routable by ν. This shows that any path of i is edge-disjoint from any path of j, or ν is a multicast infra concentrator. $\qquad\square$

Ofman proposed to use a connector as a hyperconcentrator. Thompson (1978) observed that a hyperconcentrator does exactly what an infra-connector does in reverse. By Theorem 1.6.1, a shuffled BY^{-1} is an infra-connector, hence, BY with a reversed shuffle on the output side can be used as the hyperconcentrator in Ofman's construction (see Fig. 4.3.3).

Figure 4.3.4 illustrates the connection of inputs 1, 3, 4, 6, 7 to outputs 1, 2, 3, 4, 5.

Offman's construction of a rearrangeable multicast network is to use a Beneš network B as a hyperconcentrator to connect all k live inputs to outputs 1 to k. $B \circ BY^{-1}$ then yields a copy network. Attaching another Beneš network at the end as a rearrangeable point-to-point network, a rearrangeable multicast network $B \circ BY^{-1} \circ B$ is obtained. Since the cost of B is $4N \log_2 N$, and BY^{-1}

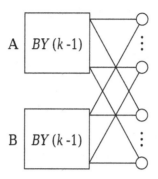

Figure 4.3.3: A recursive construction of $BY(k)$

is half of B, the total cost is $10N \log_2 N$. Note that $O(N \log N)$ is also the lower bound for multicast rearrangeable network, since it is for rearrangeable point-to-point network. Thompson noted that the first Beneš network can be replaced by BY and $2N \log_2 N$ is saved from the cost.

Lea (1988) proposed to discard the hyperconcentrator and to replace the BY^{-1} by what he called a *consecutive spreader* which serves the same function of a copy network. However, the consecutive spreader works only if the live inputs form a consecutive set $[1, k]$ for some k. The function of a hyperconcentrator is to turn an arbitrary set of live inputs to such a set.

The inverse banyan network used as a multicast infra-connector can be self-routed in $O(\log N)$ time. The banyan network used as a hyperconcentrator can also be self-routed in $O(\log N)$ time since the order-preserving property transforms the concentrator into a connector. The Beneš network used as a connector can be routed in $O(N \log N)$ time, which becomes the time to route the Ofman–Thompson network. It is not advisable to use this network for dynamic traffic since if the request from input i is blocked in the hyperconcentrator, all requests from input $i' > i$ may have to be rearranged. The number of rearrangements is thus of order $O(N)$.

As the bottleneck for routing is in the connector part, Lee (1988) suggested the use of a self-routing Batcher-(shuffled) banyan network, which is rearrangeable as discussed in Sec. 3.4, to replace the Beneš network as the connector. While the cost of the network is increased to $O(N(\log N)^2)$ and the depth to $O((\log N)^2)$, the self routing takes only $O((\log N)^2)$ time.

Nassimi and Sahni (1982) used a divide-and-conquer principle to design a multicast rearrangeable network. Let M be a parameter which divides N.

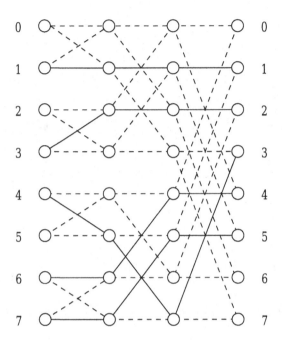

Figure 4.3.4: BY as a hyperconcentrator

Divide the N outputs into M groups of N/M each, and use a multicast rear-rangeable (N/m)-network to route the requests of each group. To guarantee that all inputs requested by a group are presented as inputs of the (N/M)-network, an $(N, N/m)$-concentrator is concatenated to the (N/m)-network. There is also a stage to distribute the N inputs into the M concentrators. The whole network is shown in Fig. 4.3.5.

The $(N, N/M)$-concentrator can be obtained from a hyperconcentrator by deleting all but N/M outputs. The multicast RNB (N/M)-network is constructed by recursion (use X_{NM} if $N \leq M$). By setting $k = \log N / \log M$, the network has $O(k \log N)$ stages with $O(kN^{1+1/k} \log N)$ cost. With $N^{1+\frac{1}{k}}$ processors, the routing requires $O(k \log N)$ time.

Again, this network does not handle the dynamic traffic well, since to accommodate a new blocked request in the $(N, N/M)$-hyperconcentrator may require the rearrangement of $O(N/M)$ connections.

Yang–Masson (1995) designed a multicast connector for dynamic traffic.

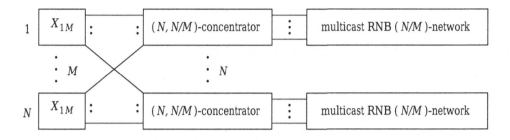

Figure 4.3.5: Nassimi–Sahni's multicast RNB network

They used a ring of M, $M \geq \max\{N_1, N_2\}$ cells to do the multicast and sandwiched this ring between an $N_1 \times M$ connector and an $M \times N_2$ connector as shown in Fig. 4.3.6.

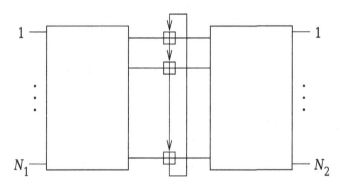

Figure 4.3.6: The Yang–Masson ring sandwich network

Partition a ring into sections S_i, $i = 1, \ldots, N_1$. A f-cast request is broken into f 1-cast requests. When a new request (i, j) comes in, check if S_i has an empty cell, route i to that cell. If not, expand S_i to the next cell and route i to it. If that cell already stores an input i', search an empty cell in $S_{i'}$ for i' as before, and if necessary, expand $S_{i'}$. Iterate this process until the search is done. If k inputs (including i) have new addresses in the ring, then $k\varphi$ rearrangements are needed in the second connector, where φ is the number of rearrangements required for one blocked request. In the worst case, $k = O(N_1)$. Using a queuing model, Yang–Masson showed that $E(k) = N_1/(M - N_2)$, where E is the expectation function. The cost of this network is clearly

$O(N \log N)$.

4.4   A Mixed Nonblocking Requirement

Hwang and Lin (1995) considered a situation where $C(n, m, r)$ is used normally for point-to-point traffic, but occasionally for 2-cast traffic. The requirement on the point-to-point traffic is SNB, while requirement on the 2-cast traffic is reduced to RNB as a side condition. Call this mixed requirement as the *HL(2)-nonblocking requirement*. They conjectured

The HL-conjecture 1. $C(n, m, r)$ satisfies the HL(2)-nonblocking requirement for $m \geq 2n$.

By Theorem 2.1.1, $m \geq 2n - 1$ is sufficient for point-to-point SNB. By Theorem 4.3.1, $m \geq 2n$ is sufficient for 2-cast traffic. However, these two results together do not imply the HL-conjecture 1 since the respective routings to achieve these two results are not compatible. The proof of Theorem 4.3.1 assumes every 2-cast connection splits at the input crossbar (unless the 2-cast is degenerated, i.e., the two outputs are on the same crossbar). This implies that each 2-cast call consumes two outputs of its input crossbar. On the other hand it is of essence to the proof of Theorem 2.1.1 that each request consumes only one output of its input crossbar. We extend Hwang-Lin's problem from 2-cast to f-cast and state a general result. The definition of HL(2)-nonblocking requirement is also extended to HL(f)-nonblocking requirement, while HL-nonblocking requirement means good for all f.

Lemma 4.4.1. *Suppose $C(n_1, r_1, m, n_2, r_2)$ is both point-to-point SNB and f-cast RNB under model 1. Then it satisfies the HL(f)-nonblocking requirement.*

Proof. Each multicast call is routed with no stage-1 fan-out. Since each request, point-to-point or multicast, consumes one output of its input switch, the sufficient condition to guarantee SNB for pure point-to-point traffic holds also for the multicast traffic.   □

Corollary 4.4.2. *Suppose $C(n_1, r_1, m', n_2, r_2)$ is f-cast RNB under model 1. Then $C(n_1, r_1, m, n_2, r_2)$, $m = \max\{m', n_1 + n_2 - 1\}$, satisfies the HL($f$)-nonblocking requirement.*

In particular, we apply Theorems 4.1.10 and 4.3.3 to obtain

Corollary 4.4.3. $C(n_1, r_1, N_2 - \lceil (N_2 + 1 - n_1)/n_2 \rceil + 1, n_2, r_2)$, *with $N_2 > n_1$, $n_2 \geq 2$, satisfies the HL(f)-nonblocking requirement.*

Proof. It was shown in the proof of Theorem 4.1.10 that

$$N_2 - \lceil (N_2 + 1 - n_1)/n_2 \rceil + 1 \geq n_1 + (n_2 - 1)\lfloor (N_2 + 1 - n_1)/n_2 \rfloor \geq n_1 + n_2 - 1.$$

□

Corollary 4.4.4. $C(n_1, r_1, \max\{[N_2(n_2 - 1)]^{1/2} + n_1 - n_2 + 1, n_2, r_2\})$ *with* $r_2 \geq 4$, *satisfies the HL(f)-nonblocking requirement.*

Proof. For $r_2 \geq 4$,

$$[N_2(n_2 - 1)]^{1/2} + n_1 - n_2 + 1 > 2(n_2 - 1) + n_1 - n_2 + 1 = n_1 + n_2 - 1.$$

□

By Lemma 4.4.1, the following conjecture implies HL-conjecture 1. *HL-conjecture 2.* $C(n, m, r)$ is 2-cast RNB under model 1 if $m \geq 2n$.

Du and Ngo (2002) extended the HL-conjecture 2 to symmetric 3-stage Clos network.

HLDN-conjecture. $C(n_1, r_1, m, n_2, r_2)$ is 2-cast RNB under model 1 with $m \geq n_1 + n_2$ and $n_1 > n_2$.

That the condition $n_1 \geq n_2$ cannot be dropped can be seen from the example of $C(2, 3, 5, 3, 4)$ with the following six 2-cast requests:$(1,\{1,2\}),(1,\{3,4\})$, $(2,\{1,3\}),(2,\{2,4\}),(3,\{1,4\}),(3,\{2,3\})$. Since every pair of requests share either the input crossbar or an output crossbar, each request must be routed through a distinct middle crossbar. But $n_1 + n_2 = 2 + 3 = 5 < 6$. Also note that $C(3, 3, 5, 3, 4)$ with the above six 2-cast requests is not RNB, showing that the condition $m \geq n_1 + n_2$ cannot be reduced to $m \geq n_1 + n_2 - 1$.

Du and Ngo proved

Theorem 4.4.5. *The HLDN conjecture holds for $n_2 = 2, 3$.*

Proof. Treat a degenerate 2-cast request as two 1-cast requests. Add enough fictitious output crossbars to maximize the number of 2-cast requests, i.e., add a fictitious 2-cast request for each idle input, connect each 1-cast request to 2-cast by adding a fictitious output crossbar to the request (ignore these connections at the end). Denote each 2-cast request by its output-crossbar pair. Define a graph G by taking the 2-cast requests as its vertices, and an edge (u, v) exists if the two requests share either the input crossbar or an output crossbar. Then the maximum degree of G is

$$n_1 - 1 + 2(n_2 - 1) = n_1 + 2n_2 - 3.$$

By Vizing's Theorem, G can be $(n_1 + 2n_2 - 2)$-colored, in particular, it can be $(n_1 + 2n - 3)$-colored if G does not contain a $(n_1 + 2n_2 - 2)$-clique.

For $n_2 = 2$, $n_1 + 2n_2 - 2 = n_1 + n_2$. Associate each color with a distinct middle crossbar, and route a request according to its color.

For $n_2 = 3$, then $n_1 + 2n_2 - 3 = n_1 + n_2$. Therefore it suffices to prove that G does not contain a $(n_1 + n_2 + 1)$-clique, or a $(n_1 + 4)$-clique.

Suppose to the contrary that there exists a $(n_1 + 4)$-clique θ. Since $n_1 + 3$ is the maximum degree in G, a vertex v in θ has no edge outside of θ. Thus the n-clique including v and the other $n_1 - 1$ requests sharing the input crossbar with v is a subgraph in θ, and this holds for every $v \in \theta$. Therefore $n_1/(n_1+4)$, n_1 is either 2 or 4. Since $n_1 \geq n_2 - 3$, n_1 must be 4. Thus θ is a 8-clique. Removing the two disjoint 4-cliques reduce θ to the complete bipartite graph $K_{4,4}$. But θ' is the line graph of a graph R with r_2' vertices (including fictitious output crossbars). Since θ' is 4-regular R must be 3-regular. So the number of edges in θ' is $3r_2'/2 = 8$ which has no integral solution for r_2, an absurdity. $\square$

Fu and Hwang (2003) proved

Theorem 4.4.6. *The HLDN conjecture is true for $r_1 = 4$ and $n_1 = n_2$.*

Proof. For each input crossbar I_i, $1 \leq i \leq 4$, construct a graph G_i by taking the 2-cast requests as vertices, and an edge (u, v) exists if the two requests share an output switch. Again, we assume that each input generates a 2-cast request by adding fictitious output crossbar, if necessary. Then G_i has n vertices.

Define $G_o = G_1 \cup G_3$, $G_e = G_2 \cup G_4$. We prove that there exists a perfect matching between G_o and G_e.

Define a bipartite graph G with G_o and G_e as the two parts and an edge between node $a = \{x, y\} \in G_o$ and $b = \{u, v\} \in G_e$ if $\{x, y\} \cap \{u, v\} = \varnothing$. Let $N(K)$ denote the set of neighbors of a set K. By Hall's Theorem on systems of distinct representatives (1935), it suffices to prove that for every $K \subseteq V(G_o)$, $|N(K)| \geq |K|$.

Suppose to the contrary that there exists a set $K \subseteq V(G_o)$ such that $|N(K)| < |K| = k$. Let $\bar{N}(K) = V(G_1)\backslash N(K)$. Then $k + \bar{N}(K) \geq 2n + 1$. Without loss of generality, assume $k \geq n + 1$. Let node $\{x, y\} \in \bar{N}(K)$. Then every node in $\bar{N}(K)$ contains either x or y. Since x or y can appear at most n times, there exists a node in $\bar{N}(K)$ containing neither x nor y. Let this node be $\{u, v\}$ where $\{u, v\} \cup \{x, y\} = \varnothing$. Then every node in K contains either u or v. Namely, every node in K is of the form $\{x, y\} \times \{u, v\}$. Note that the total number of appearances of x, y, u, v in K is at least $2k$. Hence the total number of appearances x, y, u, v in $\bar{N}(K)$ is at most $4n - 2k < 2(2n+1-k) \leq 2|\bar{N}(K)|$.

This implies the existence of a node a in $\bar{N}(K)$ containing at most one of x, y, u, v. Suppose a contains x only. Then there exists a node b in K which does not contain x (or x would appear $2n + 1$ times in K). Thus a and b are not neighbors, contradicting the definition of $\bar{N}(K)$.

Route each edge in the perfect matching through a distinct middle crossbar. Then $m = 2n$ suffices. $\qquad\qquad\square$

Hwang, Liaw and Tong (2003) considered model 1 for general f-cast. Let M denote the number of multicast requests. They proved

Theorem 4.4.7. $C(n_1, r_1, n_2, r_2, m)$ *is multicast RNB under model 1 for $M \leq m$ if and only if $m \geq \max\{n_1, n_2\}$.*

Proof. Route each multi-cast request through a distinct middle crossbar. Then each multi-cast request consumes only one output of its input crossbar. Therefore the necessary and sufficient condition in Theorem 3.1.2 remains true here. $\qquad\qquad\square$

Corollary 4.4.8. $C(n_1, r_1, m, n_2, r_2)$ *satisfies the HL-nonblocking requirement for $M \leq m$ if and only if $m \geq \min\{n_1 + n_2 - 1, N_1, N_2\}$.*

The $M \leq m$ condition can be dropped in some special case.

Corollary 4.4.9. $C(n_1, r_1, n_1 + n_2, n_2, r_2)$ *satisfies the HL-nonblocking requirement if $n_1 \geq kn_2$ for some integer k and $r_2 \leq k + 3$. If $n_1 > n_2$, then $C(n_1, r_1, n_2, r_2, n_1 + n_2 - 1)$ satisfies the HL-nonblocking requirement for $r_2 \leq 4$.*

Proof. $M \leq \lfloor r_2 n_2 / 2 \rfloor \leq \lfloor (k+3) n_2 / 2 \rfloor \leq \lfloor (n_1 + 3n_2)/2 \rfloor \leq n_1 + n_2$. $\qquad\square$

Corollary 4.4.9 proves the HLDN conjecture for $n_1 \geq n_2$ and $r_2 \leq 4$. Instead of controlling the total number of multicast requests, we can also control the maximum number k of multicast requests an input crossbar can have. Note that in Theorems 4.1.12 and 4.3.3, if k is imposed, then n_1 in the condition on m can be replaced by k. Thus Corollary 4.4.8 becomes

Theorem 4.4.10. $C(n_1, r_1, \max\{n_1 + n_2 - 1, k + f(n_2 - 1)\}, n_2, r_2)$ *and* $C(n_1, r_1, \max\{n_1 + n_2 - 1, [N_2(n_2 - 1)]^{1/2} + k - n_2 + 1\})$ *satisfy the HL-nonblocking requirement if each input crossbar can have at most k f-cast requests.*

References

Bassalygo, L. A., & Pinsker, M. S. 1980. Asymptotically optimal networks for generalized rearrangeable switching and generalized switching without rearrangement. *Problemy Peredachi Informatsii*, **16**, 94–98.

Blum, M., Karp, R. M., Papadimitriou, C. H., Vornberger, O., & Yannakakis, M. 1981. The complexity of superconcentrators. *Info. Proc. Lett.*, **13**, 164–167.

Dolev, D., Dwork, C., Pippenger, N., & Wigderson, A. 1983. Superconcentrators, generalizers and generatized connectors with limited depth. *ACM Symp. Thy. Comput.*, **15**, 42–51.

Du, D. Z., Hwang, F. K., & Richards, G. W. 1985. A problem of lines and intersections with an application to switching networks. *Pages 151–164 of: Ann. of Disc. Math. 126*. Amsterdam: North Holland.

Feldman, P., Friedman, J., & Pippenger, N. 1986. Nonblocking networks. *ACM Symp. Thy. Comput.*, **18**, 247–254.

Feldman, P., Friedman, J., & Pippenger, N. 1988. Wide-sense nonblocking networks. *SIAM J. Disc. Math.*, **1**, 158–173.

Garey, X. X., Hwang, F. K., & Richards, G. W. 1988. Asymptotic results from partial concentrators. *IEEE Trans. Commun.*, **36**, 214–217.

Giacomazzi, P., & Trecordi, V. 1995. A study of nonblocking multicast switching networks. *IEEE Trans. Commun.*, **43**, 1163–1168.

Hall, P. 1935. On representatives of subsets. *J. London Math. Soc.*, **10**, 26–30.

Hu, X. D., & Hwang, F. K. 1992. An improved upper bound for the subarray partial concentrator. *Disc. Appl. Math.*, **37/38**, 341–346.

Hwang, F. K. 1972. Rearrangeability of multiconnection three-stage networks. *Networks*, **2**, 301–306.

Hwang, F. K. 1979. Three-stage multiconnection networks which are nonblocking in the wide sense. *Bell Syst. Tech. J.*, **58**, 2183–2187.

Hwang, F. K., & Jajszczyk, A. 1986. On nonblocking multiconnection networks. *IEEE Trans. Commun.*, **34**, 1038–1041.

Hwang, F. K. 2003. A survey of nonblocking multicast three-stage Clos networks. *IEEE Commun. Mag. Oct.*, 34–37.

Hwang, F. K. 2003. A unifying approach to determine the necessary and sufficient conditions for nonblocking multicast Clos networks. *IEEE Trans. Commun.*, to appear.

Hwang, F. K., & Liaw, S. C. 2000. *On nonblocking multicast 3-stage Clos networks. IEEE/ACM Trans. Network.*, **8**, 535–539.

Hwang, F. K., Liaw, S. C., & Tong, L. D. 2003. Strictly nonblocking 3-stage Clos networks with some rearrangeable multicast capability. *IEEE Trans. Commun.*, **51**, 1765–1767.

Hwang, F. K., & Lin, B. C. 2003. Wide-sense nonblocking multicast $\log_2(N, m, p)$ networks. *IEEE Trans. Commun.*, **51**, 1730–1735.

Hwang, F. K., & Richards, G. W. 1991. Using combinatorial designs to construct partial concentrators. *IEEE Trans. Commun.*, **39**, 1141–1146.

Hwang, F. K., & Richards, G. W. 1992. The capacity of the subarray partial concentrators. *Disc. Appl. Math.*, **39**, 231–240.

Jacobsen, S. B. 1988. The rearrangement process in a two-stage broadcast switching network. *IEEE Trans. Commun.*, **36**, 484–491.

Kabacinski, W., & Danilewicz, G. 2002. Wide-sense and strict sense nonblocking operation of multicast multi-$\log_2 N$ switching networks. *IEEE Trans. Commun.*, **50**, 1025–1036.

Kirkpatrick, D. G., Klawe, M., & Pippenger, N. 1985. Some graph-coloring theorems with applications to generalized connection networks. *SIAM J. Alg. Disc. Methods*, **6**, 576–582.

Kufta, R. W., & Vacroux, A. G. 1983. Multiple stage switching with fan-out capacities. *Proc. Comput. Network. Symp.*, Silver Spring, 89–96.

Lea, C.-T. 1988. A new broadcast switching network. *IEEE Trans. Commun.*, **10**, 1128–1137.

Lee, T. T. 1988. Nonblocking copy networks for multicast packet switching. *IEEE J. Selected Areas Commun.*, **6**, 1455–1467.

Masson, G. M. 1977. Binomial switching networks for concentration and distribution. *IEEE Trans. Commun.*, **25**, 873–883.

Masson, G. M., & Jordan, Jr., B. W. 1972. Generalized multi-stage connection networks. *Networks*, **2**, 191–209.

Nassimi, D., & Sahni, S. 1982. Parallel permutation and sorting algorithms and a new generalized connection network. *J. ACM*, **29**, 642–667.

Ofman, Y. P. 1965. A universal automaton. *Trans. Moscow Math. Soc.*, **14**, 200–215.

Pinsker, M. 1973. On the complexity of a concentrator. *Proc. 7th ITC*, Stockholm, Sweden, 318/1–318/4.

Pippenger, N. 1973. *The complexity of switching networks*. MIT Res. Lab. Elec. Tech. Rep. 487.

Richards, G. W., & Hwang, F. K. 1985. A two-stage rearrangeable broadcast switching network. *IEEE Trans. Commun.*, **33**, 1025–1035.

Thompson, C. D. 1978. Generalized connection network for parallel processor interconnection. *IEEE Trans. Comput.*, **27**, 1119–1124.

West, D. B., & Banerjee, P. Partial matching in degree restricted bipartite graphs. Preprint.

Yang, Y., & Masson, G. M. 1991. Nonblocking broadcast switching networks. *IEEE Trans. Comput.*, **9**, 1005–1015.

Yang, Y., & Masson, G. M. 1995. Broadcast ring switching networks. *IEEE Trans. Comput.*, **44**, 1169–1180.

Yang, Y., & Masson, G. M. 1996. The necessary conditions for Clos-type nonblocking multicast networks. *Proc. 10th Inter. Para. Proc. Symp.*, Honolulu, Hawaii, 789–795.

Chapter 5. Multirate Networks

The need of a multirate network comes from the desire to integrate multi-media transmissions such as audio, data, image and video into one switching networks. As different media require a broad range of bandwidths (even the same media can generate requests requiring different bandwidths), each request is associated with its required rate of bandwidth. A link in the switching network now has a *capacity* and can carry as many requests as desired as long as the sum of their rates does not exceed the capacity of the link. Multirate can also be called *k-rate* if the number of distinct rates is specified to be k.

There are two basic multirate models: *discrete* and *continuous*. The discrete model assumes that there is a finite number of distinct rates. Suppose the smallest rate b divides all other rates. Then a link with capacity c can be decomposed into $\lfloor c/b \rfloor$ *channels* where a channel has the exact capacity to carry a request of the smallest rate. Then we can talk about a request as a q-channel request, and a link as an f-channel link. We call this the channel model. Note that a q-channel request must be routed in one path. If a q-channel request can be routed as q 1-channel requests, then it is no longer a genuine multirate model, but a 1-rate model where the only rate is 1.

For 3-stage Clos network, there are only 2-stage of links plus the inputs and outputs. We use f_i, $i = 1, 2$, to denote the capacity of stage-i links, f_0 for inputs and f_0' for outputs. For multistage networks, we consider a symmetric model by having f_0 as the capacity for inputs and outputs, and f_i for links in shell i. Typically, we assume $f_0 \leq f_1 \leq f_2 \leq \cdots$ to reflect the need of faster internal links, which have the same effect on nonblocking as adding internal links. In most cases only 1-level speed up is considered, i.e., $f_i = f_1$ for all $i \geq 1$. The case all f_i are equal is called f-uniform.

For the continuous model, it is customary to assume that each internal link has capacity 1 and normalize the rates into a number w, $b \leq w \leq B$, where $0 < b \leq B \leq 1$ are bounds of w. Each external link is assumed to have capacity β, $B \leq \beta$, i.e., it can generate any number of requests in a frame as long as the sum of rates of these requests does not exceed β. We call this the $\beta[b, B]$ model (which can also be used for discrete multirate). We use $\beta(b, B]$,

$\beta[b, B)$, $\beta(b, B)$ to exclude b or B, and omit β if $\beta = 1$. In practice, β is usually less than 1, representing a 1-level speed-up.

A request from input switch X to output switch Z will be denoted by either (X, Z, q) for q-channel, or (X, Z, w) for normalized rate w. We will also refer to them as a q-*request* or a w-*request*. For multirate model, the definition of co-inputs (co-outputs) of a request (x, z) should be extended to include $x(z)$ itself since $x(z)$ can generate other requests taking away links from connecting the current request. However, the current request itself should not be counted as a connection engaged by a co-input (co-output).

5.1   1-Rate Model

A degenerate multirate case is the 1-rate case where every request has a rate w, $0 < w \leq 1$. Note that $w = 1$ corresponds to the classical model discussed in the previous three chapters. Define $f = \lfloor c/w \rfloor$ for a link with capacity c. Then the link capacity can be represented in terms of the number f of requests it can carry. Suppose an input link can generate f_1 requests. Clearly, the nonblocking property of the network depends on f_1 only, i.e., it doesn't matter what is the value w as long as it yields f_1. Furthermore, the rates can even vary in a small range as long as f_1 is invariant. We extend the definition of 1-rate to include the case, which is a special case of the channel model where all requests are unit-requests. If furthermore, f is uniform over all links, then we call it the 1-rate f-uniform model.

Since the theory of the classical model is well established, it is profitable to ask how much of it can be extended to the multirate model. It is natural to start this exploration from the simplest multirate model, the 1-rate model. Recently, Chen, Hwang and Zhu (2004) proved

Theorem 5.1.1. *An acyclic network is SNB for the 1-rate f-uniform model if and only if it is SNB for the classical model.*

Proof. We will treat the network as a directed graph G with the crossbars as nodes. We also add a node at the end of an input or output links to make G a legitimate graph. If every link has f channels, we represent the network by (G, f).

(i) (G, f) *SNB* $\Rightarrow$ $(G, 1)$ *SNB*.

Suppose s is a blocking state of $(G, 1)$. Let s' be the state of (G, f) obtained by duplicating each path of s f times. Define $s'(e)$ to be the number of paths containing e. Then $s'(e) = fs(e)$. Since s is a blocking state, there exists input i and output o not contained in any path, and any path from i to o contains a link e with $s(e) = 1$. Then neither i nor o is contained in any path in s', and every path from i to o contains a link e with $s'(e) = f$. Hence s' is a blocking state in (G, f).

(ii) $(G, 1)$ *SNB* $\Rightarrow (G, f)$ *SNB*.

We first prove (ii) for $f = 2$. Let s be a blocking state of $(G, 2)$ for input i and output o. Let G_s be the subgraph of G consisting of the paths in s. Since the capacity of a link is 2, a link in G is contained in at most two paths in s. A link in G will be drawn k times, $k = 0, 1, 2$, in G_s if it is contained in k paths of G_s (thus a link in G_s has capacity 1).

We 2-color the links in G_s. For each pair of duplicated link, color them differently. The remaining links will be colored one path or one cycle at a time, while a link is considered undirected in such a path or cycle. The rule of coloring is:

(1) For a path, start at an end of it; for a cycle, start at any link. Color the link arbitrarily except that if both i and o have a path, then the link from i must have the same color as the link from o.

(2) Proceed coloring links on the path or cycle in a fixed direction. Suppose link x was just colored and link y is to be colored. Then x and y are incident to the same vertex v. If both x and y are in-links, or both are out-links of v, y is colored differently from x. If one is in-link and the other is out-link, y is colored same as x.

Note that even if i and o are on the same path, rule 2 guarantees them to receive the same color. The fact that rule 2 would not lead to contradiction in coloring a link is due to the acyclic property of G.

Since for every vertex in G_s, the number of in-links equals to the number of out-links, the above two rules guarantee that for a given color, the number of in-links also equals to the number of out-links; and this number is upper bounded by one. Assign the links to two copies of G by colors. Then the links in each copy constitute a set of path (not necessarily among the original set of paths in s), hence each set is a state of $(G, 1)$. One of them, say state s', contain neither i nor o, and every path from i to o in s' contains a link e with $s'(e) = 1$. For otherwise, then there exists a path from i to o in s which contains no link e with $s(e) = 2$, a contradiction to the fact that (i, o) is blocked in s. Hence s' is a blocking state for (i, o) in $(G, 1)$.

For the general $f \geq 2$ case, a theorem of Little, Tutte and Younger (1988), which is a variation of the Manger Theorem, says that if G is acyclic, then a state s in (G, f) can be partitioned into states $s_1, \cdots, s_f$ of $(G, 1)$ such that $s(e) = \sum_{j=1}^{f} s_j(e)$ for every link e. However, this theorem does not guarantee the existence of a s_j which contains neither i nor o. Suppose s_j does not contain i and s'_j does not contain o. Combine the paths in s_j and s'_j to obtain a state in $(G, 2)$. Then apply the $f = 2$ result.   □

For RNB, Chung-Ross (1991) proved

Theorem 5.1.2. *An MIN is 1-rate RNB if it is RNB for the classical model.*

Proof. Suppose that an external link has capacity f. Then the frame graph with inputs and outputs (not switches) as vertices has maximum degree f and hence can be f-edge-colored. Edges of the same color represent a subframe where each external link generates only one request; hence can be routed as the classical model. Route each subframe independently. A network link may have to carry f connections, one from each subframe; but that is within the capacity.   □

The other direction of Theorem 5.1.2 remains an open problem.

Corollary 5.1.3. *The d-nary Cantor network $[X_{1m}, B_N^d, X_{m1}]$ is m-rate rearrangeable.*

Proof. B_N^d is rearrangeable for the classical model. So the m rates can be routed by dedicating each copy of B_N^d to a fixed rate.   □

For WSNB, Fishburn-Hwang-Du-Gao (1997) showed that the classical model does not imply the 1-rate model, not even for 3-stage Clos networks.

Theorem 5.1.4. *Suppose $C(n, m, r)$ is WSNB under algorithm A for the f-uniform model if $m \geq m(n, r)$, where $m(n, r)$ satisfies $m(kn, r) \leq k \cdot m(n, r)$. Then $C(n, m, r)$ is WSNB for the f'-uniform model if f divides f'.*

Proof. Consider the 1-rate f'-uniform model. Treat a link as a union of $k = f'/f$ sublinks each with capacity f. Decompose each middle switches into k switches with sublinks. Then $C(n, m, r)$ under f' is transformed into $C(kn, km, r)$ under f, which is WSNB since

$$km \geq k \cdot m(n, r) \geq m(kn, r).$$

□

Fishburn–Hwang–Du–Gao also applied Theorem 5.1.4 to obtain a rare exact result for WSNB.

Theorem 5.1.5. $C(n, m, 2)$ *is WSNB for the 1-rate f-uniform model for any* $f \geq 2$ *if and only if $m \geq m^o = \lceil 3n/2 \rceil$.*

Proof. Beneš proved that $C(n, m, 2)$ is WSNB for the classical model if and only if $m \geq m^o = \lfloor 3n/2 \rfloor$ in Corollary 2.1.4. Note that $m^o(n, r) = \lfloor 3n/2 \rfloor$ does not satisfy the condition $m^o(kn, r) \leq k \cdot m^o(n, r)$. Hence we cannot conclude that $\lfloor 3n/2 \rfloor$ suffices for the f-uniform model. However $m(n, r) = \lceil 3n/2 \rceil$ does satisfy the condition, and is sufficient for WSNB for the classical model ($f = 1$). Since $f = 1$ divides any other f, $\lceil 3n/2 \rceil$ suffices for the f-uniform model.

The proof for the "only if" part is somewhat complicated and omitted. □

Corollary 2.1.4 and Theorem 5.1.5 together imply that WSNB for the classical model does not imply WSNB for the 1-rate f-uniform model. Fishburn–Hwang–Du–Gao conjectured that the converse is true and gave some support to the conjecture.

To obtain an implication from f' to f, we have to set conditions on the routing algorithm A. A is called *inert* if two requests from the same input switch to the same output switch with no connection deleted in between are routed through the same middle switch whenever capacity allows. Examples of inert algorithms are "minimum index" and "packing".

Theorem 5.1.6. *Suppose $f < f'$ and A is an inert algorithm. Then $C(n, m, r)$ is WSNB under A for f' implies it is so for f.*

Proof. Consider the f' model and assume that requests always come in by groups of $q = f'/f$ identical copies. Then $C(n, m, r)$ is still WSNB under A for such types of requests. But any sequence of requests for f can be translated into this type of requests for f' under A. Furthermore, an f'-link can always be treated as q f-links. Theorem 5.1.6 follows. □

In the rest of this section, we consider the 1-rate model but the capacity of a link can vary. If particular, the capacity of an input link is often assumed to be smaller than that of an internal link to achieve good nonblocking effect.

An $n \times m$ crossbar each of whose inlets can carry f requests and each outlet f' requests is called a *digital symmetrical matrix* (DSM), which will be simply referred to as a switch. (In the original definition of DSM, each internal link consists of v channels each can carry f' requests. For mathematical theory, the parameter v can be eliminated by setting $f' = vf'$.) Jajszczyk

(1983) gave necessary and sufficient conditions for a $C(n, m, r)$ with DSM to be SNB and RNB. Hwang–Yeh (1998) extended it to $C(n_1, r_1, m, n_2, r_2)$. They also allowed the links at different stages to have different f-values. Let $C(n_1, r_1, m, n_2, r_2; f_0, f_1, f_2, f'_0)$ denote a $C(n_1, r_1, m, n_2, r_2)$ such that the capacities of the input link, the stage-1 link, the stage-2 link and the output link are f_0, f_1, f_2 and f'_0, respectively.

Define $R_1 = \min\{n_1 f_0, N_2 f'_0\}$ and $R_2 = \min\{n_2 f'_0, N_1 f_0\}$.

Theorem 5.1.7. $C(n_1, r_1, m, n_2, r_2; f_0, f_1, f_2, f'_0)$ *is SNB if and only if* $m \geq m^o$, *where*

$$m^o = \left\lfloor \frac{R_1 - 1}{f_1} \right\rfloor + \left\lfloor \frac{R_2 - 1}{f_2} \right\rfloor + 1.$$

Proof. Without loss of generality, assume the new request is from an input switch X to an output switch Z. Then $X(Z)$ has no access to a middle switch Y if and only if either all f_1 channels of the XY-link, or all f_2 channels of the YZ-link are busy. But the co-inputs can engage all f_1 inlets of at most $\lfloor (R_1 - 1)/f_1 \rfloor$ middle DSMs, and the co-outputs at most $\lfloor (R_2 - 1)/f_2 \rfloor$. Hence m^o suffices.

On the other hand, it is easily verified that the co-inputs can fully engage $\lfloor (R_1 - 1)/f_1 \rfloor$ middle DSMs, the co-outputs can fully engage $\lfloor (R_2 - 1)/f_2 \rfloor$, and these two sets of switches can be disjoint. Hence m^o is necessary. $\square$

Corollary 5.1.8. $C(n, m, r; f_0, f_1)$ *is SNB if and only if* $m \geq m^o$, *where*

$$m^o = 2 \left\lfloor \frac{n f_0 - 1}{f_1} \right\rfloor + 1.$$

If $f_1 = f_0$, *then* $m^o = 2n - 1$.

Theorem 5.1.9. $C(n_1, r_1, m, n_2, r_2; f_0, f_1, f_2, f'_0)$ *is RNB if and only if* $m \geq m^o$, *where*

$$m^o = \max \left\{ \left\lceil \frac{R_1}{f_1} \right\rceil, \left\lceil \frac{R_2}{f_2} \right\rceil \right\}.$$

Proof. Necessity is trivial. We now prove that m^o is sufficient.

Let S be an input or output switch with $d(S)$ requests. Partition the $d(S)$ requests into $\lceil d(S)/m^o \rceil$ groups such that each group, except perhaps the last, has m^o requests. Do this for every S. Let G be the bipartite graph with the input groups and the output groups as vertices and requests between input groups and output groups as edges. Since G has maximum degree m^o, it can be m^o-edge-colored. Label the m^o middle DSMs by the m^o colors, and route all requests through the middle DSM of the same color.

An input switch has at most $\lceil R_1/m^o \rceil = f_1$ groups. Hence it sends at most f_1 requests to a given middle DSM, which can be carried by a single stage-1 link. An analogous argument goes for an output DSM. □

Corollary 5.1.10. $C(n, m, r; f_0, f_1)$ *is RNB if and only if* $m \geq m^o$, *where* $m^o = \lceil nf_0/f_1 \rceil$. *If* $f_1 = f_0$, *then* $m^o = n$.

An immediate extension is

Corollary 5.1.11. $C(n, kn, r)$ *is* k-*rate rearrangeable under the* f-*uniform model.*

Ohta (1991) proposed a repackable algorithm for $C(n, m, r)$. Let $W(X, Y, Z)$ denote the number of connections from input switch X through middle switch Y to output switch Z. Define $\underline{W}(X, Z) = \min\limits_{Y} W(X, Y, Z)$ and $\bar{W}(X, Z) = \max\limits_{Y} W(X, Y, Z)$. His algorithm is defined by the following two operations:

(i) A new request (X, Z) is always routed through a Y which achieves $\underline{W}(X, Z)$.

(ii) Suppose a connection (X, Y, Z) is removed. If Y now achieves $\underline{W}(X, Z)$, move a connection (X, Y', Z) from Y' to Y where Y' achieves $\bar{W}(X, Z)$.

Otherwise, do nothing.

It is easily seen that for any X, Y, Y', Z,

$$W(X, Y, Z) \leq W(X, Y', Z) + 1.$$

Theorem 5.1.12. $C(n, m, r; f_1, f_0)$ *is repackable under Ohta's algorithm if* $f_0 \geq \left\lfloor \frac{nf_1 - r}{m} \right\rfloor + r$.

Proof. Consider an (X, Z) request. Suppose Y^o achieves $\underline{W}(X, Z)$. We show the link from X to Y^o has a vacancy. Define $W(X, Y) = \sum\limits_{Z} W(X, Y, Z)$ and $W(X) = \sum\limits_{Y} W(X, Y)$. Then

$$
\begin{aligned}
W(X, Y^o) &= \sum_{Z'} W(X, Y^o, Z') \leq \sum_{Z' \neq Z} (W(X, Y, Z') + 1) \\
&\quad + W(X, Y, Z) = W(X, Y) + r - 1 \\
nf_1 - 1 &\geq W(X) = \sum_{Y} W(X, Y) \geq mW(X, Y^o) - (m-1)(r-1).
\end{aligned}
$$

Therefore
$$W(X, Y^o) \leq \left\lfloor \frac{nf_1 - r}{m} \right\rfloor + r - 1 < f_0 .$$

Similarly, we can show the link from Y^o to Z has a vacancy. Hence Y^o can carry (X, Z). □

Ohta stated Theorem 5.1.12 as a necessary and sufficient condition. He claimed necessity by giving an example that $W(X, Y^o) = W(X, Y) + r - 1$ for every $Y \neq Y^o$. In this example

$$W(X, Y, Z) = (nf_1 - r)/m \quad \text{for all } Y,$$

$$W(X, Y^o, Z') = 1 \text{ and } W(X, Y, Z') = 0 \quad \text{for all } Y' \neq Y^o \text{ and } Z' \neq Z.$$

However, for this example to exist, $nf_1 - r$ must be non-negative and a multiple of m. These conditions are not met by all parameters. Therefore, there exist $C(n, m, r; f_1, f_0)$ such that the condition in Theorem 5.1.12 is necessary, but not for every $C(n, m, r; f_1, f_0)$; in fact, not even for most.

5.2   The IMF Model

Suppose we relax the requirement that a request must be routed through one path. Then all requests between an input switch and an output switch can be combined, and the sum of their rates is termed the *demand* of the pair. Thus the notions of individual request and individual input (output) become moot. Suppose we also allow each internal link to have its own capacity. Then the multirate rearrangeability of a switching network can be studied through a multicommodity flow model; and if the network is under a discrete model, then the flow problem is integral (all parameters are integers).

Varma–Chalasani (1993) first introduced the integral multicommodity flow (IMF) model to study the 3-stage Clos network by allowing each link to have its own capacity. Note that the effect of external links having individual capacities is to allow each input and each output switch have its own upper bound of number of requests it can generate (how this number is broken down into capacity per link is immaterial).

Let $K(X, Y)$ or $K(Y, Z)$ denote the capacity of link (X, Y) or (Y, Z) (Y would be omitted if the quantity is invariant in Y). Let $U(X)$ or $U(Z)$ denote the upper bound of weight of all requests on X or Z.

Theorem 5.2.1. *The 3-stage Clos network for the IMF model is rearrangeable for f_2-cast if*

$$m \geq \max \left\{ \frac{\min\{U(X)f_2, N_2\}}{K(X)} : X = X_1, \ldots, X_{r_1}, \right.$$

$$\left. \frac{\min\{U(Z), N_1\}}{K(Z)} : Z = Z_1, \ldots, Z_{r_2} \right\}.$$

Proof. The proof is similar to the proof of Theorem 4.3.1 except that we have to derive a condition for each X_i (Z_j) and select the maximum. The division by $K(X)$ or $K(Z)$ is because $K(X)$ or $K(Z)$ colors can be packed into one X- or Z-link. □

They also gave an $O(mK^2)$ routing algorithm, where $K = \sum_X K(X)$ by using the Gabow–Kariv edge-coloring algorithm (see Sec. 3.2).

Liptopoulos–Chalasani (1994) extended Theorem 5.1.12 to the IMF model.

Corollary 5.2.2. *The 3-stage Clos network for the IMF model is repackable under Ohta's algorithm if*

$$\mathbf{K}(X, Y) \geq \left\lfloor \frac{U(X) - r_2}{m} \right\rfloor + r_2 \quad \text{for all } X,$$

and

$$\mathbf{K}(Y, Z) \geq \left\lfloor \frac{U(Z) - r_1}{m} \right\rfloor + r_1 \quad \text{for all } Z.$$

They also proposed a different repacking algorithm. Define

$$V(X, Y, Z) = \min \{K(X, Y) - W(X, Y), \ K(Y, Z) - W(Y, Z)\}.$$

Their algorithm is obtained from Ohta's algorithm by replacing $W(X, Y, Z)$ with $V(X, Y, Z)$. However, no proof of repackability is known.

Elmallah–Culberson (1995) studied the 3-stage Clos network under the IMF model. In their set-up, a 3-stage Clos network is viewed as a digraph without external links. The demand $D(X, Z)$ between every input node X and every output node Z is specified. Each link ℓ has a capacity C_ℓ. This flow network will be denoted by $\text{IMF}(r_1, m, r_2; C, D)$ where C and D are the capacity-set and the demand-set. The focus of the flow problem is not to determine how large m should be so that D can be routed, but rather for the given network, what condition should D satisfy for routability.

A *cut* is a set L of links which separates the nodes into two parts A and B such that every link in L is from A to B. Because of the topology of the 3-stage Clos network, either A contains no output node, or B no input node. The demand $D(L)$ of the cut is the sum of $D(X, Z)$ where X and Z are in different parts. The capacity $C(L)$ of the cuts is the sum of C_ℓ for $\ell \in L$. Clearly, $D(L) \le C(L)$ is a necessary condition for the routability of $D(L)$. The question is whether $D(L) \le C(L)$ for all L is also sufficient. We call this the *CD-condition*.

Elmallah–Culberson proved

Theorem 5.2.3. *The CD-condition is sufficient for the routability of IMF* $(r_1, m, r_2; C, D)$ *if $m \le 2$.*

Proof. The $m = 1$ case is trivial. We prove for $m = 2$.

A (g, f)-factor of a graph G is a subgraph H such that $g(v) \le d_H(v) \le f(v)$ for every v in H. Heinrich–Hall–Kirkpatrick–Lin (1990) proved that a bipartite graph G has a (g, f)-factor if and only if for every subset S of vertices

$$\sum_{x \ne S} \left(g(x) - d_{G \setminus S}(x) \right)^+ \le f(S),$$

where $y^+ = \max\{y, 0\}$ and $f(S) = \sum_{v \in S} f(v)$.

Let G be the frame graph with the set X of input nodes and the set Z of output nodes. Let $Y = \{y_1, y_2\}$ be the set of middle nodes (middle DSMs). Set $f(v) = c(v, y_1)$ and $g(v) = d_G(v) - c(v, y_2)$. Then the demands are routable if a (g, f)-factor of G exists since the factor can be routed by y_1, and the residual demands by y_2.

Suppose to the contrary that there exists a set S such that

$$\sum_{x \notin S} \left(g(x) - d_{G \setminus S}(x) \right)^+ > f(S). \qquad (*)$$

By assuming S is minimal, it can be argued that either $S \subseteq X$ or $S \subseteq Z$. Assume the former. Then every $x \in X \setminus S$ does not contribute to the left-hand side of the inequality since

$$g(x) - d_{G \setminus S}(x) = d_G(x) - c(x, y_2) - d_{G \setminus S}(x) = -c(x, y_2) \le 0.$$

Define $Z' = \{z \in Z : g(z) - d_{G \setminus S}(z) > 0\}$. Then

$$f(S) < \sum_{x \notin S} \left(g(x) - d_{G \setminus S}(x) \right)^{+} = \sum_{z \in Z'} \left(g(z) - d_{G \setminus S}(z) \right)$$

$$= \sum_{z \in Z'} \left(d_G(z) - c(z, y_2) - d_{G \setminus S}(z) \right)$$

$$= d(S, Z') - C(Z', y_2)$$

implies

$$d(S, Z') > C(Z', y_2) + C(S, y_1) \geq C(S, Z')$$

since y_1 can carry at most $c(S, y_1)$, and y_2 at most $c(Z', y_2)$ between S and Z', a contradiction to the CD-condition.   □

Elmallah–Culberson also proved

Theorem 5.2.4. *The CD-condition is sufficient to route IMF* $(r_1, m, r_2; C, D)$ *if* $r_1 \leq 2$, $c(x, y)$ *is either 0 or* c, *and* $c(z, y) = c$ *for all* $x \in X$, $y \in Y$ *and* $z \in Z$.

Proof. The case $r_1 = 1$ is trivial. Assume $r_1 = 2$ with $X = \{x_1, x_2\}$. Combine the middle nodes with both $c(x_1, y)$ and $c(x_2, y) > 0$ into a node y_s, and combine all other middle nodes into a node y_d. Apply Theorem 5.2.3 on y_s and y_d. It suffices to give an algorithm to split the connections carried by y_s and y_d back to the y_s'.

The splitting in y_d is simple. Decompose y_d into two disjoint groups of y_s', Y_1 and Y_2, with Y_1 connecting to x_1 only and Y_2 to x_2. Let R_i denote the set of connections from x_i, $i = 1, 2$, which are carried by y_d. We can partition R_i into groups of c (except one group may contain less than c) and assign each group to a node in Y_i.

The subnetwork involving y_s is a 3-stage Clos network. Since splitting is allowed in IMF routing, we may assume that all requests are unit-requests. Let G_s denote the frame graph on the connections carried by y_s, and let Δ be the maximum degree of G_s. Then G_s can be Δ-edge-colored. Since G_s can be routed by y_s, $|\{y_s\}| \geq \Delta / c$. By routing edges of c colors through one y in y_s, the job is done.   □

Elmallah–Culberson showed that with three middle nodes, to determine whether a demand set is routable is NP-complete even with uniform capacity and uniform demand. It is also difficult to extend Theorem 5.2.4 to $r_1 = 3$

as the splitting problem for y_d, the node which combines all middle nodes not connecting to all three input nodes, becomes messy.

Fingerhut–Suri–Turner (1997) considered a network model where each node $v \in V$ has an incoming capacity $K'(v)$ and an outgoing capacity $K(v)$, each arc $(u, v) \in E$ has a capacity $K(u, v)$ and a length $\ell(u, v) = \ell(v, u)$. The cost of an edge (u, v) is

$$C(u, v) = K(u, v)\ell(u, v)$$

and the cost of a network is simply $\sum_{(u,v) \in E} C(u, v)$. The problem is to find a minimum-cost network which can route any set $\{D(u, v)\}$ of demands satisfying $\sum_v D(u, v) \leq K(u)$ and $\sum_u D(u, v) \leq K'(v)$ for all u, v.

Define $K = \sum_{v \in V} K(v)$, $K' = \sum_{v \in V} K'(v)$ and

$$L = \sum_{u \in V} \sum_{v \in V} K(u)K'(v)C(u, v) .$$

In the *symmetric* version $K(v) = K'(v)$ and $K(u, v) = K(v, u)$ for all $u, v \in V$. In the *balanced* version $K = K'$ and $K(u, v) = K(v, u)$ for all $u, v \in V$. Fingerhut–Suri–Turner showed that even the symmetric version is an NP-hard problem.

Define a double star $S(v_k, v_l)$ by the conditions:

$$K(v_i, v_\ell) = K(v_i) \quad \text{for all } i \neq k, \ell,$$

$$K(v_k, v_i) = K'(v_i) \quad \text{for all } i \neq k, \ell,$$

$$K(v_k, v_\ell) = K(v_k) + K'(v_\ell),$$

$$K(v_\ell, v_k) = K',$$

$$K(v_i, v_j) = 0, \qquad \text{otherwise}.$$

Figure 5.2.1 illustrates a double star. Note that $K(v_k, v_\ell)$ is split into two parts for easier algebraic maneuvering later.

Theorem 5.2.5. *The cheapest double star has cost at most $(K+2K')L/(KK')$.*

Proof. Let DS denote a multiset of double stars with $S(v_k, v_\ell)$ appearing $K(v_k)K'(v_\ell)$ times over all $v_k, v_\ell \in V$. Then

$$|\text{DS}| = \sum_{v_k \in V} \sum_{v_\ell \in V} K(v_k)K'(v_\ell) = \sum_{v_k \in V} K(v_k) \sum_{v_\ell \in V} K'(v_\ell) = KK' .$$

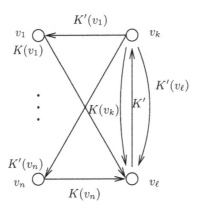

Figure 5.2.1: A double star

The arc (v_k, v_ℓ) contributes a cost of $\ell(v_k, v_\ell)K(v_k)K'(v_i)K'(v_\ell)$ in $S(v_k, v_i)$ for all $i \neq k$, a cost of $\ell(v_k, v_\ell)K(v_i)K'(v_\ell)K(v_k)$ in $S(v_i, v_\ell)$ for all $i \neq \ell$, and an extra cost of $\ell(v_k, v_\ell)K(v_\ell)K'(v_k)K'$ in $S(v_\ell, v_k)$. Adding up, the total contribution is

$$\ell(v_k, v_\ell)\left[\sum_{i \neq k} K(v_k)K'(v_i)K'(v_\ell) + \sum_{i \neq \ell} K(v_i)K'(v_\ell)K(v_k) + K(v_\ell)K'(v_k)K'\right]$$

$$\leq \ell(v_k, v_\ell)\left[K(v_k)K'(v_\ell)K' + K'(v_\ell)K(v_k)K + K(v_\ell)K'(v_k)K'\right].$$

Summing over all v_k, v_ℓ, the average cost of a double star in DS is bounded by $(K + 2K')L/KK'$. Theorem 5.2.5 follows immediately. □

A *star network* is obtained from the double-star network by merging v_k and v_ℓ.

Corollary 5.2.6. *The cheapest star costs at most $2L/K$ for the symmetric version.*

Proof. By deleting the extra cost of $\ell(v_k, v_\ell) K(v_\ell) K'(v_k)K$ in $S(v_\ell, v_k)$, and setting $K = K'$. □

Fingerhut–Suri–Turner then gave a lower bound of the cost of a network, star or not star.

Theorem 5.2.7. *Suppose ℓ is a metric, $K \geq K'$ and $K(v_i)$, $K'(v_i)$ are integers for all $v_i \in V$. Then there exists a demand set such that the cost of any network is at least L/K.*

Proof. Let G be a bipartite graph with $K(v)$ and $K'(v)$ copies of node v on the "source" part and the "sink" part, respectively. Then there are K sources, K' sinks, and the total length of all edges is L.

There are $\binom{K}{K'}$ ways of choosing K' sources and $K'!$ ways to match these K' sources with the K' sinks. So there are $K!/(K-K')!$ maximum matchings. Each arc is counted

$$\binom{K-1}{K'-1}(K'-1)! = \frac{(K-1)!}{(K-K')!}$$

times in these matchings. Hence the total length of these matchings is $L(k-1)!/(K-K')!$. The average length of a matching is thus L/K.

Let M denote the longest maximum matching. Then its length is at least L/K. Consider M as a demand set. If ν is an SNB network for M, then

$$K_\nu(u,v) \geq \min\{K(u), K'(v)\}.$$

Furthermore, ℓ being a metric implies

$$\ell(u,v) \leq \min_t\{\ell(u,t) + \ell(t,v)\}.$$

Hence the cost of ν

$$
\begin{aligned}
C_\nu &= \sum_{(u,v)\in E} K_\nu(u,v)\ell(u,v) \\[2mm]
&\geq \sum_{(u,v)\in E} \min\{K(u), K'(v)\}\ell(u,v) \\[2mm]
&\geq \text{length of the matching } M \geq L/K.
\end{aligned}
$$

$\square$

Corollary 5.2.8. *The cost of the cheapest double star is at most $2 + K/K'$ times over that of the cheapest network. The ratio is at most 3 for the balanced version and at most 2 for the symmetric version.*

Note that a star is always cheaper than the corresponding double star. The reason of getting into double stars at all is a technical one. For the unbalanced model, a way has not been found to express the cost of a star which can be easily compared with the lower bound of Theorem 5.2.7, as Corollary 5.2.8 did for double stars.

5.3 The Channel Model

Niestegge (1988) first formally studied the channel model. Let Q denote the upper bound of a request(in terms of number of channels). Chung-Ross (1991) proved Theorem 5.3.1 for $C(n, m, r; f)$.

Theorem 5.3.1. $C(n_1, r_1, m, n_2, r_2; f_0, f_1, f_2, f_0')$ is SNB if and only if $m \geq m^o$, where

$$m^o = \left\lfloor \frac{n_1 f_0 - Q}{f_1 - Q + 1} \right\rfloor + \left\lfloor \frac{n_2 f_0' - Q}{f_2 - Q + 1} \right\rfloor + 1.$$

Proof. Consider a request (X, Z, q). Let Y be a middle switch. Then the XY-link is inadequate for the request if and only if it has already used up $f_1 - q + 1$ channels. Since the inputs on X, not counting the current request, can claim at most $n_1 f_0 - q$ channels for connection, at most $\lfloor (n_1 f_0 - q)/(f_1 - q + 1) \rfloor$ middle switches have inadequate links to X. Similarly, at most $\lfloor (n_2 f_0' - q)/(f_2 - q + 1) \rfloor$ middle switches have inadequate links to Z. Hence

$$\left\lfloor \frac{n_1 f_0 - q}{f_1 - q + 1} \right\rfloor + \left\lfloor \frac{n_2 f_0' - q}{f_2 - q + 1} \right\rfloor + 1$$

middle switches suffices. The maximum of the above expression is obtained at $q = Q$, giving m^o. It is easy to construct an example such that the worst scenario can occur. □

Kabacinski (1995) considered $C(n, m, r; f_0, f_1, v)$ which can be obtained from $C(n, m, r; f_0, f_1)$ by replacing each internal link with v copies. The reason that this network is different from $C(n, m, r; f_0, v f_1)$ for the multirate case is that the v links cannot share (sharing implies different paths) their capacities in routing a q-request in the former network. (For the 1-rate-class model, i.e., $q = 1$, the need of sharing does not exist; so the two networks are the same.) Kabacinski proved (he stated the result as WSNB)

Corollary 5.3.2. $C(n, m, r; f_0, f_1, v)$ is SNB if and only if $m \geq m^o$, where

$$m^o = 2 \left\lfloor \frac{n f_0 - Q}{v(f_1 - Q + 1)} \right\rfloor + 1.$$

He also considered the case that the f_1 channels in an internal link are linearly ordered and the q channels carrying a q-channel-request must be consecutive in a link. The reason of this requirement is that the consecutiveness makes it easy to reconstruct the request (put the packets of the request back to the original order) when using the result for multislot assignment in an ATM network.

Corollary 5.3.3. *Suppose a q-channel-request must be carried by q consecutive channels in a link. Then $C(n, m, r; f_0, f_1, v)$ is SNB if and only if $m \geq m^o$, where*

$$m^o = 2 \left\lfloor \frac{nf_0 - Q}{v \lfloor f_1/Q \rfloor} \right\rfloor + 1 .$$

Proof. In the worst scenario, the $nf_0 - Q$ co-inputs can generate $nf_0 - Q$ 1-channel requests with $\lfloor f_1/Q \rfloor$ of them scattered in a middle switch with a constant distance of $Q - 1$ channels apart. Then there are no Q consecutive channels available in this switch to carry a Q-request.   □

Chung–Ross considered the d-nary Cantor network and proved the following result for the f-uniform channel model. We will allow the external link to have f_0 channels and the internal link to have f_1 channels. An internal link is q-saturated if it already carries a load of at least $f_1 - q + 1$.

Theorem 5.3.4. *The d-nary Cantor network $[X_{1m}, B_n^d, X_{m1}]$ is SNB for the (Q, f_0, f_1) model if and only if $m \geq m^o$, where*

$$m^o = 2 \left\lfloor \frac{d\ g(Q)}{N} \right\rfloor + 1 ,$$

$$g(q) = \sum_{k=2}^{n} d^{n-k} \left\{ \left\lfloor \frac{d^{k-1} f_0 - q}{f_1 - q + 1} \right\rfloor - \left\lfloor \frac{d^{k-2} f_0 - q}{f_1 - q + 1} \right\rfloor \right\} .$$

Proof. Consider an (x, z, q) request. Let $CG(x, z)$ denote the channel graph. Let ℓ_k, $2 \leq k \leq n$, denote the number of stage-k saturated links in $CG(x, z)$. Whose preceding links are not saturated(to avoid overlapping in counting blocked middle switches). Note that the stage-1 links always have enough capacity since $q < f_1$. Since B_n^d has the buddy property, there are d^{k-1} inputs which can reach the same set of stage-k nodes as x. These inputs generate a maximum of $d^{k-1} f_0 - q$ traffic. Hence the number of saturated links in the first k stages is at most

$$\sum_{i=2}^{k} \ell_i \leq \left\lfloor \frac{d^{k-1} f_0 - q}{f_1 - q + 1} \right\rfloor \quad \text{for} \quad 2 \leq k \leq n . \tag{$*$}$$

Since a stage-k saturated link blocks d^{n-k} middle switches, $L = \sum_{k=2}^{n} \ell_k d^{n-k}$ middle switches are blocked from the input side. To maximize L subject to the constraints $(*)$, it is easily verified that the greedy solution

$$\ell_2^o = \left\lfloor \frac{df_0 - q}{f_1 - q + 1} \right\rfloor,$$

$$\ell_k^o = \left\lfloor \frac{d^{k-1}f_0 - q}{f_1 - q + 1} \right\rfloor - \left\lfloor \frac{d^{k-2}f_0 - q}{f_1 - q + 1} \right\rfloor \quad \text{for} \quad 3 \le k \le n$$

does it. Therefore $L = g(q)$ and it can be verified that $g(q)$ is maximized at $q = Q$. The same argument works for the output side. In the worst scenario the two sets of $g(Q)$ middle switches are disjoint. Since each B_n^d has N/d middle switches, the $2g(Q)$ middle switches can exhaust all the middle switches of at most $\lfloor 2d\, g(Q)/N \rfloor\ B_n^d$. Hence one more B_n^d guarantees the existence of a middle switch not in any of the $g(Q)$ sets. Theorem 5.3.4 follows by noting that the worst scenario can be realized. □

Recently, Hwang, He and Wang (2003) extended Theorem 5.3.4 to the $\log_d(N, k, m)$ network.

Theorem 5.3.5. *Consider the* (Q, f_0, f_1) *channel model with* $d^{\lfloor \frac{n-1}{2} \rfloor} f_0 \ge f_1 + 1$. *Then* $\log_d(N, 0, m)$ *is multirate strictly nonblocking if and only if* $m \ge m^o$, *where*

$$m^o = \left\lfloor \frac{d^{\lfloor \frac{n-1}{2} \rfloor} f_0 - Q}{f_1 - Q + 1} \right\rfloor + \left\lfloor \frac{d^{\lceil \frac{n-1}{2} \rceil} f_0}{f_1 - Q + 1} \right\rfloor + 1.$$

Proof. Suppose the new request is (i, o, q). Call an internal link q-saturated if it always carries a load exceeding $f_1 - q$ (hence cannot carry a new q-request). Note that the channel graph between i and o is just a single path. Consider the link l in the channel graph between stage $\lfloor n/2 \rfloor$ and $\lfloor n/2 \rfloor + 1$. Then $d^{\lfloor (n-1)/2 \rfloor}$ inputs and $d^{\lceil (n-1)/2 \rceil}$ outputs can reach l. Since each input and output can be involved in requests with a total weight f_0, these inputs (outputs) occupy $d^{\lfloor \frac{n-1}{2} \rfloor} f_0\ (d^{\lceil \frac{n-1}{2} \rceil} f_0)$ channels. But q channels must be subtracted from it since the new request (i, o, q) implies that i and o have q channels unused. Thus the total weight from the $d^{\lfloor \frac{n-1}{2} \rfloor}$ inputs can saturate l at most

$$\left\lfloor \frac{d^{\lfloor \frac{n-1}{2} \rfloor} f_0 - q}{f_1 - q + 1} \right\rfloor$$

times. Similarly, the total weight from the $d^{\lceil \frac{n-1}{2} \rceil}$ outputs can saturate l at most

$$\left\lfloor \frac{d^{\lceil \frac{n-1}{2} \rceil} f_0 - q}{f_1 - q + 1} \right\rfloor$$

times. In the worst case, these two saturated sets go to disjoint copies of $BY^{-1}(n,0)$. Thus

$$\left\lfloor \frac{d^{\lfloor \frac{n-1}{2} \rfloor} f_0 - q}{f_1 - q + 1} \right\rfloor + \left\lceil \frac{d^{\lceil \frac{n-1}{2} \rceil} f_0 - q}{f_1 - q + 1} \right\rceil + 1$$

copies of $BY_d^{-1}(n,0)$ suffice to carry the new request. Noticing that $d^{\lfloor \frac{n-1}{2} \rfloor} f_0 \geq f_1 + 1$, then is maximized when $q = Q$. Thus the sufficiency part of Theorem 5.3.5 is proved.

It can be verified that the worst case constructed in the sufficiency proof can happen. Thus the condition in Theorem 5.3.5 is also necessary. $\qquad \square$

Next consider the general k case. Define

$$g_k(q) = \sum_{j=1}^{k} \frac{1}{d^j} \left\{ \left\lfloor \frac{d^j f_0 - q}{f_1 - q + 1} \right\rfloor - \left\lfloor \frac{d^{j-1} f_0 - q}{f_1 - q + 1} \right\rfloor \right\} \quad for \ 0 \leq k \leq n - 1$$

Theorem 5.3.6. *Consider the (Q, f_0, f_1) channel model with $d^{\lfloor \frac{n-1}{2} \rfloor} f_0 \geq f_1 + 1$. Then $\log_d(N, k, m)$ is multirate strictly nonblocking for $0 \leq k \leq n - 1$ if and only if $m \geq m^o$, where*

$$m^o = \max_{1 \leq q \leq Q} \left\{ \lfloor g_k(q) + \frac{d^{\lfloor \frac{n+k-1}{2} \rfloor} f_0 - q - \left\lfloor \frac{d^k f_0 - q}{f_1 - q + 1} \right\rfloor (f_1 - q + 1)}{d^k (f_1 - q + 1)} \rfloor + \right.$$

$$\left. \lfloor g_k(q) + \frac{d^{\lceil \frac{n+k-1}{2} \rceil} f_0 - q - \left\lfloor \frac{d^k f_0 - q}{f_1 - q + 1} \right\rfloor (f_1 - q + 1)}{d^k (f_1 - q + 1)} \rfloor + 1 \right\}$$

Proof. The $k = 0$ was proved in Theorem 5.3.5 and the $k = n - 1$ (the Cantor network) case in Theorem 5.3.4. We prove the general k case.

The argument of Theorem 5.3.4 to prove $p \geq 2\lfloor g_{n-1}(Q) \rfloor$ is necessary and sufficient for the Cantor network to be multirate strictly nonblocking computes the total weight of requests which can reach a stage-j link, $1 \leq j \leq (n+k-1)/2$, in the (i, o) channel graph to be $d^j f_0 - q$, while a q-saturated link carries a load at least $f_1 - q + 1$. Thus at most

$$\left\lfloor \frac{d^j f_0 - q}{f_1 - q + 1} \right\rfloor$$

links in the channel graph at or before stage j can be saturated. The worst case is to assign the saturated links to as early a stage as possible since links in the early stages have more blocking power. This results in assigning

$$\lfloor \frac{d^j f_0 - q}{f_1 - q + 1} \rfloor$$

saturated links to stage j, each of which blocks $1/d^j$ copies of a $BY_d^{-1}(n, k)$. Counting also the output side, then $2\lfloor g_k(q) \rfloor$ is the number of copies of $BY_d^{-1}(n, k)$ blocked by paths intersecting the links of the (i, o) channel graph in the first or last k stages.

Finally, the total weight of requests which can reach at or before a stage-j link, $1 \leq j \leq \lfloor (n + k - 1)/2 \rfloor$, is

$$d^{\lfloor \frac{n+k-1}{2} \rfloor} f_0 - q.$$

But a total weight of

$$\lfloor \frac{d^k f_0 - q}{f_1 - q + 1} \rfloor (f_1 - q + 1)$$

was already connected in the first k stage. Therefore only the difference of there two weights can be used to saturate stage-j links for $k \leq j \leq n$. Since the $\log_d(N, k, m)$ network between these stages consists of d^k copies of $BY_d^{-1}(n - k, 0)$, we apply Theorem 5.3.5 for the number of copies of $BY_d^{-1}(n - k, 0)$ blocked, which must be divided by d^k to convert to the number of copies of $BY_d^{-1}(n, k)$ blocked.

Again, it can be verified that the worst case constructed here can happen. □

There is no positive WSNB result particular for the discrete model except those inherited from the continuous model (see next section). Tsai, Wang and Hwang (2001) gave a lower bound for r large.

Theorem 5.3.7. *$C(n, m, r)$ is not WSNB for any two rates B, b satisfying $B + b > 1$ if $m \leq \min\{\lfloor 1/b \rfloor, n\} + 2n - 3$ and r is large enough.*

Proof. Without loss of generality, we may assume that $B = 1$ and $b = 1/k$ for some integer $k \geq 2$. We consider two phases.

In phase 1, all requests have weight $1/k$ and come from the first input of each input switch (outputs will be specified when needed).

Step 1. Each input generates one request. Since r is large enough, there exists a large set I_1 of inputs, routed through the same middle switch M_1.

Step 2. Each input generates a second request. Let $k + 1$ second requests generated by I_1 go to the same output switch. This is possible since $b(k + 1) \leq nk$, the capacity of an output switch. But the total weight $(k + 1)/k$ requires at least a new middle switch M_2 other than M_1 to carry. Let $i \in I_1$ be an input whose two requests are carried by M_1 and M_2. Since r is large enough, there exists a large set I_2 of inputs whose two requests are carried by M_1 and M_2.

. . .

Let $t = \min\{k, m\}$. At the end of step $t - 1$, there exists a large set I_{t-1} of inputs whose $t - 1$ requests are all carried by $M_1, M_2, \cdots, M_{t-1}$.

Step t. Each input generates a tth request. Let $(t - 1)k + 1$ inputs in I_{t-1} all go to the same output switch. Since $(t - 1)k + 1 \leq nk$, this is possible. But the total weight $[(t - 1)k + 1]/k > t - 1$, so at least one such request must be carried by a new middle switch M_t. Since r is large enough, actually there exists a large set I_t of inputs whose k requests are carried by $\{M_1, \ldots, M_t\} \equiv M$.

In phase 2, consider only the input switches generating I_t. Let each of them generate $n - 1$ weight-1 requests going to a set of clean output switches. None of these requests can be routed through M. By Theorem 2.1.8, another set of $2n - 2$ middle switches is needed. Hence a total of $t + 2n - 2$ middle switches is needed. $\square$

The weight model given in Theorem 5.3.6 can be stated in terms of channel. Let a channel have the exact capacity to carry a b-request. Define $q = \lfloor 1/b \rfloor$. Then

Corollary 5.3.8. $C(n, m, r)$ *is not WSNB for two rates q and 1 if $m \leq$ $\min\{q, n\} + 2n - 3$ and r is large enough*

Corollary 5.3.8 applies to weights $w_1 \geq w_2 \geq \cdots \geq w_k \geq 1/2 \geq b$ with $w_k + b > 1$.

Corollary 5.3.9. *For $k \geq n$, then $m \leq 3n - 3$ is impossible for WSNB.*

While Theorem 5.1.2 confirms that a rearrangeable network for classical model remains so for 1-rate, the following example by Chung–Ross shows that it is not so for 2-rate. Consider B_2^3 with two rates $(2, 1)$ and $f = 2$. Suppose the frame (on inputs and outputs) is $\{(x_1, z_1, 2), (x_2, z_2, 2), (x_4, z_4, 2), (x_5, z_5, 2),$

$(x_6, z_3, 1), (x_7, z_6, 2), (x_9, z_3, 1)\}$. Without loss of generality, assume (x_6, z_3, y_2) is routed through Y_3. Then $(x_1, z_1, 2)$ and $(x_2, z_2, 2)$ must be routed through Y_1 and Y_2 (since z_1, z_2, z_3 are all in Z_1), while $(x_4, z_4, 2)$ and $(x_5, z_5, 2)$ must also be routed through Y_1 and Y_2 (since x_1, x_2, x_3 are all on X_1). The latter fact forces $(x_7, z_6, 2)$ to be routed through Y_3 (since z_4, z_5, z_6 are all on Z_2), which blocks $(x_9, z_3, 1)$ to go to Y_3 to share a link with $(x_6, z_3, 1)$, which it has to.

We imitate the proof of Theorem 5.4.18 to get a channel model version

Theorem 5.3.10. $C(n, m, r)$ *is rearrangeable for the* (Q, f) *model if*

$$m \geq \frac{nf - Q}{f - Q}.$$

Proof. In Theorem 5.5.1 an external link has capacity β, an internal link has capacity 1, and $w \leq B$. Replace the link capacity, be it β or 1, by f, and replace B by Q (or $w \leq B$ by $q \leq Q$). $\square$

Chung–Ross commented it would be of interest to see if an SNB network for the classical model guarantees rearrangeability for the channel model. For convenience, we refer to this question as the *Chung–Ross conjecture*. It turns out that this conjecture has since been studied in some special cases of the discrete model, which we will present in Sec. 5.5.

5.4 The Continuous Model: Nonblockingness

We assume $\beta n \geq 1 \geq \beta \geq B$ for the continuous model. All results about symmetric 3-stage Clos network can be extended to the asymmetric case by setting $n = \max\{n_1, n_2\}$, $r = \max\{r_1, r_2\}$.

Melen–Turner (1989) proved

Theorem 5.4.1. $C(n, m, r)$ *is SNB for the* $\beta[b, B]$ *model if*

$$m \geq 2 \max_{b \leq w \leq B} \left\lfloor \frac{\beta n - w}{M(w)} \right\rfloor + 1,$$

where $M(w) = \max\{1 - w + \epsilon, b\}$ *and* ϵ *is a positive number approaching zero.*

Proof. Consider a request (X, Z, w) and a middle switch Y. Link (X, Y) cannot carry this request if and only if it has already carried a load exceeding $1 - w$. On the other hand, the inputs of X can generate a load at most $\beta n - w$ carried in the network. So the number of middle switches Y such that the (X, Y) link is w-saturated is at most $\lfloor (\beta n - w)/M(w) \rfloor$ (note that such a switch carries a load exceeding b). The same argument holds for the (Y, Z) link. Thus m^o suffices. $\qquad \square$

Note that if $M(w)$ can be realized, then the sufficient condition in Theorem 5.4.1 is also necessary. We consider two such special cases. First a lemma,

Lemma 5.4.2. *Suppose that there exists an integer p such that $b \le 1/p < B$. Then there exists a combination of weights w, $b \le w \le B$, whose sum is $1 + \epsilon$.*

Proof. Take $p - 1$ weights of $1/p$ and one weight of $1/p + \epsilon$. $\qquad \square$

Corollary 5.4.3. *Suppose $b + B \le 1$ and $b \le 1/p \le B$ for some integer p. Then $C(n, m, r)$ is SNB if and only if $m \ge m^o$, where*

$$m^o = 2 \left\lceil \frac{\beta n - B}{1 - B} \right\rceil - 1.$$

Proof. $b + B \le 1$ implies $M(w) = 1 - w + \epsilon$ for $b \le w \le B$. Hence the condition in Theorem 5.4.1 becomes

$$\begin{aligned}
m &\ge 2 \max_{b \le w \le B} \left\lfloor \frac{\beta n - w}{1 - w + \epsilon} \right\rfloor + 1 \\
&= 2 \left\lfloor \frac{\beta n - B}{1 - B + \epsilon} \right\rfloor + 1 \qquad \text{since } \beta n \ge 1 \\
&= 2 \left\lceil \frac{\beta n - B}{1 - B} \right\rceil - 1.
\end{aligned}$$

To prove necessity, note that for $w = 1/p$, $M(w)$ can be realized by Lemma 5.4.2. $\qquad \square$

Suppose no such integer p exists. For $p = 1$, then this is the 1-rate model studied in Sec 5.1. By Theorem 5.1.1, $m^o = 2n - 1$.

Corollary 5.4.4. *Suppose $b + B > 1$. Then $C(n, m, r)$ is SNB if*

$$m \ge 2 \left\lceil \frac{\beta n - 1}{b} \right\rceil + 1.$$

Proof. Note that

$$M(w) = \begin{cases} 1 - w + \epsilon & \text{for } b \le w \le 1 - b, \\ b & \text{for } 1 - b < w \le B. \end{cases}$$

Hence the condition in Theorem 5.2.1 becomes

$$m \geq \max \left\{ 2 \max_{b \leq w \leq 1-b} \left\lfloor \frac{\beta n - w}{1 - w + \epsilon} \right\rfloor + 1, \, 2 \max_{1-b \leq w \leq B} \left\lfloor \frac{\beta n - w}{b} \right\rfloor + 1 \right\}$$

$$= \max \left\{ 2 \left\lfloor \frac{\beta n - 1 + b}{b + \epsilon} \right\rfloor + 1, \, 2 \left\lfloor \frac{\beta n - 1 + b}{b} \right\rfloor + 1 \right\} = 2 \left\lceil \frac{\beta n - 1}{b} \right\rceil + 1 \, .$$

□

Chung–Ross tightened the condition in Corollary 5.4.4 and showed it is also necessary. While they did it for $B = \beta = 1$, we state for more general β and B and with a different proof.

Theorem 5.4.5. $C(n, m, r)$ *is SNB for the* $\beta[b, B]$ *model satisfying* $b + B > 1 \geq \beta$ *if and only if* $m \geq m^o$, *where* $m^o = 2 \lfloor \beta/b \rfloor (n - 1) + 1$.

Proof. Each co-input and co-output can generate at most $\lfloor \beta/b \rfloor$ connections. Even if each such connection is carried by a distinct middle switch, there still exists a middle switch not carrying any of these connections, which can be used to carry the current request. Note that whether this switch carries any connection from the co-requests is immaterial, as the total load an input or output can generate is $\beta \leq 1$.

On the other hand the co-inputs and co-outputs can indeed generate these $m^o - 1$ connections, and route them through distinct middle switches if one exists. A new request with weight B cannot be routed through these switches, since $B > 1 - b$, and requires an extra switch. □

For the Cantor network with continuous $\beta[b, B]$, the situation is analogous to the 3-stage Clos network case. Namely, necessary and sufficient conditions for SNB are known only if $b + B > 1$. Chung–Ross proved the following theorem for $\beta = 1$.

Theorem 5.4.6. *The d-nary Cantor network is SNB for continuous* $\beta[b, B]$ *satisfying* $b + B > 1 \geq \beta$ *if and only if* $m \geq m^o$ *where*

$$m^o = 2 \left\lfloor \lfloor \frac{\beta}{b} \rfloor \frac{d - 1}{d} (n - 1) \right\rfloor + 1 \, .$$

Proof. Suppose the current request is from input i to output o. Again, we count only blockage from inputs and outputs other than i and o. Each input can generate at most $\lfloor \beta/b \rfloor$ requests. Consider the channel graph $CG(i, o)$ for input i and output o. Each stage-k link, $2 \leq k \leq n$, can be grabbed by $(d-1)d^{k-2}$ inputs (buddies), and each such link can reach d^{n-k} middle switches. Assuming all these involved middle switches are disjoint, the total number is

$$\lfloor \beta/b \rfloor \sum_{k=2}^{n}(d-1)d^{k-2}d^{n-k} = \lfloor \beta/b \rfloor (n-1)(d-1)d^{n-2} .$$

Since each B_n^d has N/d middle switches, the involved middle switches can fill at most

$$\left\lfloor \frac{\lfloor \beta/b \rfloor (n-1)(d-1)d^{n-2}}{N/d} \right\rfloor = \left\lfloor \lfloor \frac{\beta}{b} \rfloor (n-1)\frac{d-1}{d} \right\rfloor$$

B_n^d. Counting also the output side, then this number is multiplied by 2. Hence m^o B_n^d suffice.

To prove necessity, assume the new request has rate B. Since $B > 1 - b$, this request cannot be connected in the $m^o - 1$ B_n^d, when each input and output generates a maximum number of b-requests which are routed as described in the "sufficiency" part. $\square$

For general $\beta[b, B]$, Melen-Turner obtained the following result which, actually, started the study of the SNB multirate Cantor network.

Theorem 5.4.7. *The d-nary Cantor network is SNB for continuous $\beta[b, B]$ if*

$$m \geq \frac{2\beta}{dM(B)} \left[1 + (d-1)(n-1)\right] .$$

Proof. Using the same argument as in the proof of Theorem 5.3.4, we obtain

$$\sum_{i=2}^{k}\ell_i \leq \left\lfloor \frac{d^{k-1}\beta - w}{M(w)} \right\rfloor \leq \frac{d^{k-1}\beta - w}{M(w)} .$$

Treating ℓ_i as real numbers, then the solution maximizing L is

$$\ell_2 = \frac{d\beta - w}{M(w)} ,$$

$$\ell_k = \frac{d^{k-1}\beta - w}{M(w)} - \frac{d^{k-2}\beta - w}{M(w)} = \frac{(d-1)d^{k-2}\beta}{M(w)} \quad \text{for } 3 \leq k \leq n .$$

It follows

$$L = \sum_{k=2}^{n} \lfloor \ell_k \rfloor d^{n-k} \;\leq\; \frac{1}{M(w)} \left[(d\beta - w)d^{n-2} + \sum_{k=3}^{n} (d-1)d^{k-2}\beta d^{n-k} \right]$$

$$= \frac{d^{n-2}}{M(w)} \left[d\beta - w + (n-2)(d-1)\beta \right]$$

$$= \frac{d^{n-2}}{M(w)} \left[\beta - w + (n-1)(d-1)\beta \right]$$

$$< \frac{d^{n-2}\beta}{M(w)} \left[1 + (n-1)(d-1) \right]$$

$$= \frac{d^{n-2}\beta}{M(B)} \left[1 + (n-1)(d-1) \right].$$

By also considering the output side, and noting each B_n^d has N/d middle switches, the number of B_n^d must be at least

$$\frac{\left[2d^{n-2}\beta/M(B) \right] \left[1 + (n-1)(d-1) \right]}{N/d} = \frac{2\beta[1 + (n-1)(d-1)]}{dM(B)}.$$

$\square$

Note that B_n^d can be considered as a d-nary Cantor network with $m = 1$. Hence

Corollary 5.4.8. B_n^d *is SNB for* $\beta[b, B]$ *if*

$$B \leq \beta \leq \frac{dM(B)}{2[1 + (d-1)(n-1)]}.$$

Recently, Hwang, He and Wang (2003) obtained

Theorem 5.4.9. *Consider the continuous* $\beta[b, B]$ *model satisfying* $b + B \leq 1$, $d^{\lfloor \frac{n-1}{2} \rfloor}\beta > 1$ *and* $b \leq 1/p < B$ *for some integer* p. *Then* $\log_d(N, 0, m)$ *is strictly nonblocking if and only if*

$$m \geq \lceil \frac{d^{\lfloor \frac{n-1}{2} \rfloor}\beta - B}{1 - B} \rceil + \lceil \frac{d^{\lceil \frac{n-1}{2} \rceil}\beta - B}{1 - B} \rceil - 1$$

Proof. With an argument analogous to the proof of Theorem 5.3.5, we obtain the sufficient condition to route an (i, o, w) new request to be

$$\lfloor \frac{d^{\lfloor \frac{n-1}{2} \rfloor} \beta - w}{1 - w + \epsilon} \rfloor + \lfloor \frac{d^{\lceil \frac{n-1}{2} \rceil} \beta - w}{1 - w + \epsilon} \rfloor + 1$$

$$= \lceil \frac{d^{\lfloor \frac{n-1}{2} \rfloor} \beta - w}{1 - w} \rceil + \lceil \frac{d^{\lceil \frac{n-1}{2} \rceil} \beta - w}{1 - w} \rceil - 1$$

which is maximized at $w = B$. The necessity again follows from Lemma 5.4.2.
□

Suppose no such integer p exists and $\beta = 1$. Then the model is reduced to the 1-rate model and Theorem 5.1.1 applies.

Theorem 5.4.10. *Consider the continuous $\beta[b, B]$ model satisfying $b + B > 1$. Then $\log_d(N, 0, m)$ is strictly nonblocking if and only if*

$$m \geq 2 \lfloor \beta/b \rfloor (d^{n-1} - 1) + 1.$$

Proof. Let the new request be (i, o, w). There are d^{n-1} inputs which can reach a link in the (i, o)-channel graph (a path). Except for the input i, each of the $d^{n-1} - 1$ other inputs can generate at most $\lfloor \beta/b \rfloor$ requests. Assign each request to a distinct copy of $BY_d^{-1}(n, 0)$. Then the input side blocks at most $\lfloor \beta/b \rfloor (d^{n-1} - 1)$ copies. Similarly, the output side blocks at most an equal number of copies. Thus one extra copy suffices to carry the new request. Note that whether the extra copy carries any load from i or o is immaterial since the load cannot exceed $\beta - w$.

If the condition $b \leq 1/k < B$ is not met, say, $1/(f_0 + 1) \leq b \leq B < 1/f_0$, then every internal link can carry a maximum of f_1 connections. If further, $\beta/(f_1 + 1) \leq b \leq B \leq \beta/f_1$, then this case can be treated as the channel model and Theorem 5.3.6 applies with $Q = 1$.

On the other hand, suppose $w = B$. Then the worst case described above can happen and the new request cannot be routed through any link already carrying a load b. Hence the sufficient condition is also necessary.   □

Next we consider the $k > 0$ case. Define

$$g_k(w) = \sum_{j=1}^{k} \frac{1}{d^j} \{ \lceil \frac{d^j \beta - w}{1 - w} \rceil - \lceil \frac{d^{j-1} w}{1 - w} \rceil \} \text{ for } 1 \leq k \leq n - 1.$$

Theorem 5.4.11. *Consider the continuous $\beta[b, B]$ model satisfying $b + B \leq 1$. Then $\log_d(N, k, m)$ is strictly nonblocking if and only if*

$$
m \geq \max_{b \leq w \leq B} \Big\{ 2\lfloor g_k(w) \rfloor + \Big\lceil \frac{d^{\lfloor \frac{n+k-1}{2} \rfloor} \beta - w - (\lceil \frac{d^k \beta - w}{1-w} \rceil - 1)(1 - w)}{d^k(1 - w)} \Big\rceil
$$

$$
+ \Big\lceil \frac{d^{\lceil \frac{n+k-1}{2} \rceil} \beta - w - (\lceil \frac{d^k \beta - w}{1-w} \rceil - 1)(1 - w)}{d^k(1 - w)} \Big\rceil - 1 \Big\}
$$

Proof. Analogous to the proof of Theorem 5.3.6. □

Theorem 5.4.12. *Consider the continuous $\beta[b, B]$ model satisfying $b + B > 1$. Then $\log_d(N, k, m)$ is strictly nonblocking if and only if*

$$
m \geq 2\lfloor \frac{k(d-1)\lfloor \beta/b \rfloor}{d} \rfloor + \lfloor \frac{(d^{\lfloor \frac{n+k-1}{2} \rfloor} - 1)\lfloor \beta/b \rfloor}{d^k} \rfloor + \lfloor \frac{(d^{\lceil \frac{n+k-1}{2} \rceil} - 1)\lfloor \beta/b \rfloor}{d^k} \rfloor + 1.
$$

Proof. Each input can generate at most $\lfloor \beta/b \rfloor$ requests. Let each internal link carry at most one request. Then there are $d^j - d^{j-1}$ inputs generating $(d^j - d^{j-1})\lfloor \beta/b \rfloor$ requests to intersect a stage-j link in the (i, o)-channel graph for $1 \leq j \leq m$. Since each such intersecting path blocks $1/d^j$ copies of $BY_d^{-1}(n, m)$, they block a total of

$$
\lfloor \sum_{j=1}^{k} \frac{(d^j - d^{j-1})\lfloor \beta/b \rfloor}{d^j} \rfloor = \lfloor \frac{k(d-1)\lfloor \beta/b \rfloor}{d} \rfloor
$$

copies. Similarly, the output side blocks the same number of copies. Finally, stage $m + 1$ to stage $n - k - 1$ consists of d^k copies of $BY_d^{-1}(n - k, 0)$. We use an argument analogous to the proof of Theorem 5.3.6 to compute the number of copies blocked in these stages to be

$$
\lfloor \frac{(d^{\lfloor \frac{n+k-1}{2} \rfloor} - 1)\lfloor \beta/b \rfloor}{d^k} \rfloor + \lfloor \frac{(d^{\lceil \frac{n+k-1}{2} \rceil} - 1)\lfloor \beta/b \rfloor}{d^k} \rfloor.
$$

So one extra copy suffices to route the new request.

To prove necessity, let $w = B$ then the worst case discussed above can happen. □

When $b \to 0$ and $B \to 1$, all SNB results presented in this section say that $m^o \to \infty$. Namely, in the $(0, 1]$ model when the rates have no restriction, SNB is not a viable property. This fact urges us to look for WSNB or RNB results.

While the classic model contains few constructive results for WSNB, the multirate model proves to be a futile ground since routing a request according to its rate provides a new dimension to design algorithms. Melen–Turner opened the door with the following result (they stated for $\beta = B = 1$). We will treat βn as an integer for easier presentation. The condition $\beta n \geq 1$ is needed if βn is not an integer.

Theorem 5.4.13. $C(n, 8\beta n - 4, r)$ *is WSNB for* $\beta(0, B]$.

Proof. Route all w-requests with $w > 1/2$ in one set of m middle switches. By Corollary 5.4.3 with $b = 1/2$,

$$m \geq 2 \left\lfloor \frac{\beta n - 1}{b} \right\rfloor + 3 = 2(2\beta n - 2) + 3 = 4\beta n - 1 \quad \text{suffices.}$$

Route all w-requests with $w \leq 1/2$ in the other set of $4\beta n - 3$ middle switches. By Corollary 5.4.3,

$$m \geq 2 \left\lceil \frac{\beta n - B}{1 - B} \right\rceil - 1 = 2(2\beta n - 1) - 1 = 4\beta n - 3 \quad \text{suffices.}$$

□

Instead of using Corollary 5.4.3, we can use Theorem 5.4.5 to route all w-requests with $w > 1/2$ in $2n - 1$ middle switches. Then $8\beta n - 4$ in Theorem 5.4.13 can be changed to $4\beta n + 2n - 4$.

Gao–Hwang (1997) gave a further improvement. Define

$$f(\beta) = \lfloor \{n \lceil \beta(q + 1) - 1 \rceil - 1\}/q \rfloor.$$

Suppose all rate w satisfy $1/q \geq w > 1/(q + 1)$. Then each internal link has q channels and each external link has $\lceil \beta(q + 1) \rceil - 1$ channels. By Corollary 5.1.3, $C(n, 2f(\beta) + 1, r)$ is SNB.

Let $R(\ell)$ denote the algorithm of reserving ℓ middle switches only for large requests (which can overflow to other middle switches). We ignore the requirement on integrality for easier presentation. Define $q = \lfloor 1/B \rfloor$. Call a request *large* if $w > 1/(q + 1)$, and *small* otherwise.

Theorem 5.4.14. $C(n, m, r)$ *is WSNB under* $R(\ell)$ *for* $\beta(0, B]$ *if*

$$m \geq 2\beta n(q + 1)(Bq + B + q - 1)/q^2$$

and

$$\ell = \lceil 2\beta n(q + 1)(Bq + B - 1)/q^2 \rceil.$$

Proof. Consider the algorithm $R(\ell)$. Suppose to the contrary that a request (X, Z, w) is blocked. First assume $w > 1/(q+1)$. Then each reserved middle switch must be carrying a load of q large requests, and each other middle switch a load exceeding $1 - w \geq 1 - B$, to deny connection of the current request. Thus the total load carried exceeds

$$\frac{2\beta n(q+1)(Bq+B-1)}{q^2} \cdot \frac{q}{q+1} + \frac{2\beta n(q+1)}{q}(1-B) = 2\beta n,$$

a contradiction to the fact that X and Z can generate at most $2(\beta n - w)$ carried load.

Next assume $w \leq 1/(q+1)$. Then each nonreserved middle switch must carry a load exceeding $q/(q+1)$ to deny the connection of the current request. Thus the total load carried exceeds

$$\frac{2\beta n(q+1)}{q} \cdot \frac{q}{q+1} = 2\beta n,$$

again, a contradiction. Theorem 5.4.14 follows. $\qquad\square$

Corollary 5.4.15. $C(n, 2n\lceil 2\beta - 1\rceil - 1 + \lceil 3.75\beta n\rceil, r)$ *is WSNB for* $\beta(0, B]$.

Proof. Route requests $> 1/2(q = 1)$ through a set of $2f(\beta)+1 = 2[n\lceil 2\beta - 1\rceil - 1] + 1 = 2n\lceil 2\beta - 1\rceil - 1$ middle switches, and apply Theorem 5.4.14 with $B = 1/2$ to small requests. $\qquad\square$

For $B \leq 1/2$, then there is no reason to reserve any middle switch for large requests since there are none. So the algorithm of Corollary 5.4.15 is reduced to that of Theorem 5.4.14. For $q = 1$, the lower bound of m in Theorem 5.4.14 becomes $8B\beta n$. It has a crossover point, assuming $\beta \leq 1$, at about $B = 1/4\beta + 15/32$ when compared with $2n\lceil 2\beta - 1\rceil - 1 + 3.75\beta n$. For $\beta = 1$, this crossover point is about $23/32$. Namely, use Theorem 5.4.14 for B up to $23/32$, but beyond that, switch to Corollary 5.4.15. Then $5.75n$ middle switches suffice.

The algorithm of Corollary 5.4.15 recursively uses the $R(\ell)$ algorithm. Namely, it first divides rates according to whether $w > 1/2$ or $w \leq 1/2$. Then it further divides the second class into $1/2 \geq w > 1/3$ and $w \leq 1/3$. We can proceed to as many rounds as we like. But the next step, i.e., dividing the class $w \leq 1/3$ into $1/3 \geq w > 1/4$ and $w \leq 1/4$ reduces the number of middle switches required only by the amount $\beta n/81$.

Liptopoulos–Chalasani (1996) extended Ohta's repackable algorithm from 1-rate to multirate. Let $W(X, Y, Z)$ denote the sum of rates over all requests from X to Z carried through Y. Let Δ denote a prespecified quantity satisfying $2B \leq \Delta < \min\left\{\frac{\mathbf{K}(X)-B}{r_2-1}, \frac{\mathbf{K}(Z)-B}{r_1-1}\right\}$, (see Sec. 5.2 for notation). Algorithm (LC) consists of the following two operations:

(i) Always route an (X, Z)-request through the Y^* such that $W(X, Y^*, Z) = \min_{Y} W(X, Y, Z)$.

(ii) If a connection $(X, Y, Z; w)$ is deleted and there exists a Y' such that

$$W(X, Y, Z) - w \leq W(X, Y', Z) - \Delta,$$

then move a set of connections (X, Y', Z) with total rate ρ, $w \leq \rho \leq \Delta$, from Y' to Y (the existence of such a set is guaranteed by the condition $\Delta \geq 2B$).

Clearly, these two operations assure that

$$W(X, Y, Z) \leq W(X, Y', Z) + \Delta \quad \text{for all } X, Y, Y', Z.$$

Theorem 5.4.16. *The 3-stage Clos network for the MF model is repackable under LC if*

$$m \geq \max\left\{\max_{X} \frac{U(X) - (r_2-1)\Delta - b}{\mathbf{K}(X) - (r_2-1)\Delta - B}, \ \max_{Z} \frac{U(Z) - (r_1-1)\Delta - b}{\mathbf{K}(Z) - (r_1-1)\Delta - B}\right\}.$$

Proof. Consider the request (X, Z, w). Let Y^o minimizes $W(X, Y, Z)$. Then for all Y,

$$W(X, Y^o, Z) \leq W(X, Y, Z)$$

$$W(X, Y^o, Z') \leq W(X, Y, Z') + \Delta \quad \text{for all } Z' \neq Z.$$

Hence

$$W(X, Y^o) \leq W(X, Y) + (r_2-1)\Delta.$$

Consequently,

$$mW(X, Y^o) \leq W(X) - w + (m-1)(r_2-1)\Delta \leq U(X) - w + (m-1)(r_2-1)\Delta,$$

or

$$W(X, Y^o) \leq \frac{U(X) + (m-1)(r_2-1)\Delta - w}{m}.$$

So if

$$\mathbf{K}(X) \geq \frac{U(X) + (m-1)(r_2-1)\Delta}{m} \quad \text{for every } X,$$

or equivalently,

$$m \geq \frac{U(X) - (r_2 - 1)\Delta - w}{\mathbf{K}(X) - (r_2 - 1)\Delta - w},$$

then the (X, Y')-link can carry the request. Note that

$$m \geq \frac{U(X) - (r_2 - 1)\Delta - b}{\mathbf{K}(X) - (r_2 - 1)\Delta - B} \geq \frac{U(X) - (r_2 - 1)\Delta - w}{\mathbf{K}(X) - (r_2 - 1)\Delta - w}.$$

Theorem 5.4.16 follows by also considering the (Y', Z)-link. $\qquad\square$

Corollary 5.4.17. $C(n_1, r_1, m, n_2, r_2)$ *for* $\beta(b, B]$ *is repackable under LC if*

$$m \geq \max\left\{\frac{\beta n_1 - (r_2 - 1)\Delta - b}{1 - (r_2 - 1)\Delta - B}, \frac{\beta n_2 - (r_1 - 1)\Delta - b}{1 - (r_1 - 1)\Delta - B}\right\}.$$

Melen–Turner gave the first result on multirate rearrangeability.

Theorem 5.4.18. $C(n, m, r)$ *is rearrangeable for* $\beta[b, B]$ *if* $m \geq (\beta n - B)/(1 - B)$.

Proof. Let X denote an input or output switch, and let $W(X)$ denote the set of rates of requests involving X. Order the rates in $W(X)$ from large to small and partition them into $\lceil |W(X)|/m \rceil = g$ groups such that each group, except perhaps the last, has m members.

Construct a bipartite graph G with the input-switch groups as one part, the output switch groups as the other part, and requests as edges. By construction, G has maximum degree m and can be m-edge-colored. Let $W_{ij}(X)$ denote the rate of the edge with color j (if any) in the ith group of X. Define $W_j(X) = \sum_{i=1}^{g} W_{ij}(X)$. For any two colors j and k,

$$W_{ij}(X) \geq W_{i+1,k}(X) \quad \text{for } i = 1, \ldots, g - 1.$$

Hence

$$W_j(X) \geq W_k(X) - W_{1k}(X) \geq W_k(X) - B \quad \text{for all } k \neq j.$$

It follows

$$\beta n \geq \sum_{j=1}^{m} W_j(X) \geq mW_k(X) - (m - 1)B,$$

or

$$W_k(X) \leq \frac{\beta n}{m} + \frac{(m - 1)B}{m}.$$

By setting

$$\frac{\beta n}{m} + \frac{(m-1)B}{m} \leq 1,$$

or equivalently, $m \geq (\beta n - B)/(1 - B)$, $W_k(X)$ can be routed through one link to the kth middle switch.   □

For $B \to 1$, the lower bound of m in Theorem 5.4.18 approaches infinity. Du–Gao–Hwang–Kim (1998) removed this unpleasant fact by using a different routing algorithm to handle large requests.

We abstract the following result from Theorem 5.4.18 for easier reference later.

Corollary 5.4.19. *Suppose the requests at each input or output switch can be partitioned into g groups such that the sum of any g weights, one from each group, does not exceed one. Then $m \geq$ the maximum group size (over all input and output switches) suffices.*

Corollary 5.4.20. $C(n, 3n - 1, r)$ *is rearrangeable for the* $[b, B]$ *model.*

Proof. A w-request is large if $w > 1/2$; it is small otherwise. By Corollary 5.1.11, the large requests, all in one rate-class, can be carried by n middle switches. Then apply Theorem 5.4.18 on the small requests. With $B = 1/2$, they can be carried by another set of $2n - 1$ middle switches.   □

Du–Gao–Hwang–Kim found an even better solution. First a lemma.

A graph with weighted edges is said to be c-colorable if for each vertex the total weight of the set of edges incident to it and having the same color is at most 1. The degree of a vertex is the sum of weights of its edges.

Lemma 5.4.21. *Consider a bipartite graph G with maximum degree n and weighted edges. Suppose that all edges with weight $w > 1/q$, q an integer, are colored by a set C of $c \geq 2n$ colors. Then $\lceil (c - 2)/q - c + 2n \rceil$ extra colors together with C can color all edges with weight $w \leq 1/q$.*

Proof. Color as many small edges by C as possible. Let U denote the bipartite graph of the uncolored small edges. We prove $\Delta(U) \leq c - 2 - (c - 2n)q$.

Let X be a vertex in U and let (X, Z) denote the edge of X with a minimum weight w. Since (X, Z) cannot be colored by C,

$$W_j(X) + W_j(Z) \geq \max\{W_j(X), W_j(Z)\} > 1 - w \quad \text{for every color } j.$$

Hence

$$W(X) + W(Z) > c(1 - w).$$

But $W(Z) \leq n - w$, hence

$$W(X) > c(1 - w) - (n - w) = c - n - (c - 1)w.$$

Therefore the sum of weights of edges incident to X in U is at most

$$n - W(X) < (c - 1)w - c + 2n.$$

Divided by w, we translate weight back to number of edges. Thus

$$d_U(X) < c - 1 - (c - 2n)/w \leq c - 1 - (c - 2n)q \quad \text{for } c \geq 2n.$$

Since each weighted edge can contain at least q small weights, Lemma 5.4.21 follows. □

Theorem 5.4.22. $C(n, m, r)$ *is rearrangeable for the* $[b, B]$ *model if* $m \geq (41n - \epsilon_n)/16$, *where* $\epsilon_n = 8, 5, 6, 3$ *for* $n \equiv 0, 1, 2, 3 (mod\ 4)$, *respectively.*

Proof. We will only prove for the $n \equiv 0 (mod\ 4)$ case as the other cases are similar.

Call a request *large* if $w > 1/2$, *medium* if $1/2 \geq w > 1/4$, and *small* if $w \leq 1/4$. Consider a vertex V_i. Let ℓ_i denote the number of large requests generated by V_i, and m_i the number of medium requests. Then

$$\ell_i \leq n,$$

$$m_i \leq 3n - 2\ell_i,$$

since each input can generate a maximum of 3 medium requests, but only one if it also generates a large request. Let L denote the bipartite graph of large requests and M that of medium requests. Split M into two disjoint subgraphs M_1 and M_2 such that the degree at every vertex is evenly split (this can always be done, in fact, Sec. 3.2 gave such an algorithm). Merge M_1 with L. Then each vertex V_i in $L + M_1$ has degree at most

$$\ell_i + m_i/2 \leq 3n/2,$$

and hence can be $(3n/2)$-edge-colored. Furthermore

$$\Delta(M_2) \leq (3n - 2\ell_i)/2 \leq 3n/2.$$

Since $w \leq 1/2$ for a medium edge, M_2 can be $3n/4$ colored.

Let C denote the set of $3n/2 + 3n/4 = 9n/4$ colors which has colored all edges with weight $> 1/4$. Apply Lemma 5.4.21,

$$\left\lceil \frac{9n/4 - 2}{4} - \frac{9n}{4} + 2n \right\rceil = \left\lceil \frac{5n - 8}{16} \right\rceil$$

additional colors together with C can color all small weights. Theorem 5.4.22 is proved by noting

$$\frac{9n}{4} + \left\lceil \frac{5n - 8}{16} \right\rceil = \left\lceil \frac{41n - 8}{16} \right\rceil .$$

$\square$

Recently, Hu–Jia–Du–Hwang (2001) studied the monotone routing algorithm in counting a set of multirate requests through $C(n, k, r)$ with $k \geq n+1$. Order the requests from large weights to small weights. Suppose the current request (I, O, w) is blocked. Let x_i^j denote the total weight of requests generated by input $i \in I$ and routed through middle switch j. Then necessarily,

$$x_1^1 + x_2^1 + \cdots + x_n^1 + w > 1,$$
$$x_1^2 + x_2^2 + \cdots + x_n^2 + w > 1,$$

$$\vdots$$

$$x_1^k + x_2^k + \cdots + x_n^k + w > 1,$$
$$x_1^1 + x_1^2 + \cdots + x_1^k + w \leq 1,$$
$$x_2^1 + x_2^2 + \cdots + x_2^k + w \leq 1,$$

$$\vdots$$

$$x_n^1 + x_n^2 + \cdots + x_n^k + w \leq 1,$$

where $x_i^j(x_i^j - w) \geq 0$, meaning any positive x_i^j is at least w. Let $I(n, k)$ denote the above system of inequalities.

Theorem 5.4.23. *If $I(n, k)$ has no solution, then $C(n, 2k - 1, r)$ is multirate RNB.*

Proof. Suppose to the contrary that a request (I, O, w) is blocked in $C(n, 2k - 1, r)$. Then for each of the $2k - 1$ middle switches, either its input link or its output link is saturated. Without loss of generality, we may assume that the input links of k middle switches as saturated. Then necessarily, $I(n, k)$ has a solution. $\square$

Theorem 5.4.24. *If $I(n,k)$ has a solution, then $(k-n)/(k-1) < w \le 1/3$.*

Proof. Subtract the sum of the last n inequalities from the sum of the top k inequalities, we obtain

$$(k-1)w > k-n \text{ or } w > (k-n)/(k-1).$$

Suppose $w > 1/3$. Then each inequality can have at most two nonzero variables (not counting w) except the $(k+1)^{st}$ can have only one. Further, the number of 1-nonzero-variable inequalities among the top k ones cannot exceed that of the last n ones. Hence the number of nonzero variables in the top k inequalities and in the last n inequalities are not equal, contradicting the fact that they should be. $\qquad\square$

Corollary 5.4.25. *$I(n,k)$ has no solution if $k \ge n + \lfloor n/2 \rfloor$.*

Hu-Jia-Du-Hwang also proved $I(n,k)$ has no solution for $2 \le n \le 4$, $k \ge n+1$ and $5 \le n \le 6$, $k \ge n+2$.

Ngo and Vu (2003), slightly improving a result of Ngo (2003), obtained a sufficient condition on m which depends on r.

Theorem 5.4.26. *$C(n, 2n-1 + \lceil (r-1)/2 \rceil, r)$ is multirate RNB.*

Proof. For any pair of input switch I and output switch J, continuously combine any two weights W_i and W_j into a new weight $W_i + W_j$ if the sum does not exceed 1. At the end, there is at most one weight $\le 1/2$ for each pair (I, J). Call those weights $> 1/2$ large and those $\le 1/2$ small. Then clearly, I can have at most $2n-1$ large weights or the total weight would be over n. We prove that I has at most $r-1$ small weights.

Recall that each pair (I, J) has at most one small weight. Hence I has at most r small weights. If I has r small weights and $2n-1$ large weights. Then there exists a J carrying both a large weight and a small weight from I. Since the other $2n-2$ large weights have a sum exceeding $n-1$, the sum of the two weights in (I, J) must be less than 1 and should have been combined. If I has at most $2n-2$ large weights, we simply reclassify a small weight as a large weight (such a reclassification cannot cause blocking).

We route the large weights and small weights separately. In each routing, each weight is treated as a request as in the classical model. By the fundamental theorem of rearrangeability, $2n-1$ middle switches suffice to route the large weights and $r-1$ middle switches suffice to route the small weights. But since a small weight $\le 1/2$, we can merge two middle switch into one and the traffic carried by each link is still within the capacity 1. Hence $\lceil (r-1)/2 \rceil$ middle switches suffice for the small weights. □

Corollary 5.4.27. *For* $r \leq (n/2^{k-1}) + 1$, $m = 2n - 1 + \lceil n/2^k \rceil$ *suffices.*

They also gave the following result which can be useful for very small r.

Theorem 5.4.28. $C(n, \lceil n(r+1)/2 \rceil, r)$ *is multirate RNB.*

Further, they gave a lower bound of m.

Theorem 5.4.29. $C(n, m.r)$ *is not multirate RNB if*

$$m < \begin{cases} 5n/4, & \text{for } n \text{ even,} \\ (5n-1)/4, & \text{for } n \text{ odd.} \end{cases}$$

Proof. If suffices to prove for $r = 2$, and the proof is through an example.

Suppose there are n(even) requests of type-1:$(I_1, O_1, 0.6)$, n requests of type-2:$(I_1, O_2, 0.4)$ and $n/2$ requests of type-3:$(I_2, O_2, 1)$. Then each of the n type-1 requests requires a distinct middle switch. Suppose k of these switches are shared by type-2 requests. Then considering the requests from I_1, at least $n + (n-k)/2$ middle switches are needed. On the other hand, considering the requests from O_2, then at least $k + n/2$ middle switches are needed. Since

$$\max\{n + (n-k)/2, k + n/2\} \geq 5n/4,$$

Theorem 5.4.29 is proved. The odd n case can be similarly proved. □

In particular, for $n = 2$, Theorem 5.4.29 gave a necessary condition $m \geq 3$, which Ngo and Vu also proved to be sufficient with an intriguing argument. Thus we have

Theorem 5.4.30. $C(2, m, r)$ *is multirate RNB if and only if* $m \geq 3$.

5.5   The Discrete Model

Not many results are available for this model; most are related to the Chung-Ross conjecture (Sec. 5.3). Chung–Ross proved (they set $B = 1$):

Theorem 5.5.1. *A network which is SNB for the classical model is rearrangeable for two rates B and b satisfying $b + B = 1$.*

Proof. If $B \leq 1/2$, then Theorem 5.5.1 follows from Theorem 5.4.18. Suppose $B > 1/2$. Connect the B-requests first. Delete all paths (including the inputs and outputs on them) from the network. The residual network should still be SNB, hence rearrangeable, for the residual inputs and outputs in the classical model. By Theorem 5.1.2, it can connect all the b-requests. $\square$

Lin–Du–Hu–Xue (1999) studied the Chung–Ross conjecture under a different weight model. Let

$$w_1 > w_2 > \cdots > w_{i-1} > 1/2 \geq w_i > w_{i+1} > \cdots > w_k$$

be the k rates. Then the half channel model requires w_k divides w_j for $i \leq j \leq k - 1$, and the recursive half channel model requires w_j divides w_{j-1} for $i + 1 \leq j \leq k$.

Lemma 5.5.2. *Let L and L' be two sets of l links carrying the same set of weights in the half channel model. Then every link in L is w_k-saturated if and only if every link in L' is.*

Proof. Define

$$K_j = \begin{cases} \lfloor (1 - w_j)/w_k \rfloor, & \text{for } 1 \leq j \leq i - 1, \\ w_j/w_k, & \text{for } i \leq j \geq k - 1, \\ \lfloor 1/w_k \rfloor, & \text{for } j = k. \end{cases}$$

Suppose the set of weights contains n_j w_j for $1 \leq j \leq i - 1$. Then the total load carried by l w_k-saturated links must be

$$w_k \sum_{j=1}^{i-1} n_j K_j + (l - \sum_{j=1}^{i-1} n_j) K_k.$$

Lemma 5.5.2 follows immediately. $\square$

Lin–Du–Hu–Xue proved

Theorem 5.5.3. *$C(n, 2n-1, r)$ is rearrangeable for the recursive half channel model.*

Proof. If $w_1 \leq 1/2$, then the recursive half channel model degenerates into the channel model. Theorem 5.5.3 follows from Theorem 5.3.10 by setting

$Q = f/2$. Therefore we may assume that $w_1 > 1/2$. Suppose there are k distinct weights. Theorem 5.5.3 is proved by induction on k. The $k = 1$ case follows from Corollary 5.1.8. We consider general k.

If $w_k > 1/2$, then Corollary 5.5.3 again follows from Corollary 5.1.8. So assume $w_k \leq 1/2$. By induction, we can connect all requests whose rates are not w_k. At last, we start to connect the w_k-requests.

Suppose a request (X, Z, w_k) is blocked. Then each of the $2n - 1$ middle switches Y must either have a saturated (X, Y)-link, or a saturated (Y, Z)-link. Without loss of generality, assume that n middle switches have saturated (X, Y)-links. By Lemma 5.5.2, the n inputs of X must all be saturated. This contradicts the assumption that X generates a new request (X, Z, w_k).   □

Lin–Du–Hu–Xue showed that better results can be obtained for 2-rate and 3-rate (recall that Corollary 5.1.11 already says that $C(n, 2n, r)$ is 2-rate rearrangeable).

Theorem 5.5.4. $C(n, 2n - 1, r)$ *is multirate rearrangeable for the discrete model if either $B \leq 1/2$ or b is the only rate $< 1/2$.*

Proof. If $B \leq 1/2$, apply Theorem 5.4.1 to conclude

$$m \geq (n - 1/2)/(1 - 1/2) = 2n - 1 \text{ suffices.}$$

If all rates, other than possibly b, $> 1/2$, apply Theorem 5.5.3.   □

Corollary 5.5.5. $C(n, 2n - 1, r)$ *is 2-rate rearrangeable.*

Theorem 5.5.6. $C(n, \lceil 7n/3 \rceil, r)$ *is 3-rate rearrangeable for the discrete model.*

Proof. Let (B, w, b) denote the three rates. Suppose $B \leq 1/2$. Then by Theorem 5.5.1, $2n$ middle switches suffice. If $w > 1/2$ or $b > 1/3$, then the three rates constitute only two rate classes. By Corollary 5.1.11, again $2n$ middle switches suffice. Suppose $b \leq 1/6$. Then Theorem 5.5.6 follows from Lemma 5.4.21 by first routing B-requests and w-requests separately, each with n middle switches, and then set $q = 6$.

Therefore we may assume $B > 1/2 \geq w$ and $1/3 \geq b > 1/6$. We consider three subcases:

Case (i). $w \leq 1/3$. Let $G(w, b)$ be the frame graph consisting of all w-requests and b-requests. Since each input (output) can generate at most five requests, $\Delta(G(w, b)) \leq 5n$. By Konig's Theorem, $G(w, b)$ can be $5n$-colored.

Let S be a set of n colors. Then requests of the other $4n$ colors can be routed through $\lceil 4n/3 \rceil$ middle switches since each internal link can carry three requests.

Let $G(B,S)$ be the frame graph combining B-requests with the requests in S. If $B + b > 1$, then each input can generate at most one request. Hence $\Delta(G(B,S)) \leq n$ and n middle switches suffice. Assume $B + b \leq 1$. Partition the requests at each input and output switch into two groups. If $B + w > 1$, then the w-requests join the B-requests in a group; if $B + w \leq 1$, then the w-requests join the b-requests. In either case, each input and output can generate at most one request in each group. Hence the maximum size of a group is n. It is also easily seen that the sum of two weights, one from each group, is at most one. By Corollary 5.4.19, n middle switches can route $G(B,S)$.

Case (ii). $w > 1/3$ and $b > 1/4$. Then $\Delta G(w,b) \leq 3n$ and $G(w,b)$ can be $3n$-colored. Requests of $2n$ colors can be routed through n middle switches since $w \leq 1/2$, and $G(B,S)$ can be routed through n middle switches just as in Case (i).

Case (iii). $w > 1/3$, $b \leq 1/4$ and $B + w > 1$. Use Corollary 5.5.5 to route the B and w requests in $2n - 1$ middle switches.

There are four types of b-saturated links: B, ww, w, ϕ, meaning the link contains the specified B or w-requests and a maximum number of b requests. Since a type-B outlink must come from a type-B inlink, there is no contribution to an expansion of the number of saturated links. Among the other three types, a necessary condition for any conversion of saturated links is that there must be an equal number k of type-ww and type-ϕ links converted to $2k$ type-w links. Again, no expansion is gained since $k + k = 2k$.

Case (iv). $w > 1/3$, $b \leq 1/4$ and $B + w \leq 1$. X has at most $2n$ B- or w-requests, including at most n B-requests. Therefore, the B-requests and the w-requests can be partitioned into two groups, with at most n requests in a group, such that all B-requests are in the same group. Since the sum of two weights, one from each group, does not exceed one, by Corollary 5.4.19, n-middle switches suffice to route all B and w requests.

By Theorem 5.1.2, another set of n middle switches can route all b-requests.

$\square$

Note that the $b \leq 1/4$ condition is not used in Case (iv).

References

Chen, W. R., Hwang, F. K., & Zhu, X. 2004 Equivalence of the one-rate model to the classical model on strictly nonblocking switching networks. *SIAM J. Disc. Math.*, **17**, 446–452.

Chung, S.-P., & Ross, K. W. 1991. On nonblocking interconnection networks. *SIAM J. Comput.*, **20**, 726–736.

Du, D. Z., Gao, B., Hwang, F. K., & Kim, J. H. 1998. On multirate rearrangeable Clos networks. *SIAM J. Comput.*, **28**, 463–470.

Elmallah, E. S., & Culberson, J. C. 1995. Multicommodity flows in simple multistage networks. *Networks*, **25**, 19–30.

Fingerhut, J. A., Suri, S., & Turner, J. S. 1997. Designing least cost nonblocking broadband networks. *J. Alg.*, **18**, 287–309.

Fishburn, P. C., Hwang, F. K., Du, D. Z., & Gao, B. 1997. On 1-rate wide-sense nonblocking for 3-stage Clos networks. *Disc. Appl. Math.*, **78**, 75–87.

Gao, B., & Hwang, F. K. 1997. Wide-sense nonblocking for multirate 3-stage Clos networks. *Theor. Comput. Sci.*, **182**, 171–182.

Heinrich, K., Hell, P., Kirkpatrick, D., & Liu, G. 1990. A simple existence criterion for (gcf)-factors. *Disc. Math.*, **85**, 313–317.

Hu, X. D., Jia, X. H., Du, D. Z. & Hwang, F. K. 2001. Monotone Routing in multirate rearrangeable networks. *J. Para. Distr. Comput.*, **61**, 1382–1388.

Hwang, F. K., He, Y. & Wang, Y. 2003. Strictly nonblocking multirate $\log_d(N, m, p)$ networks. *preprint*.

Hwang, F. K., & Yeh, H. G. 1998. On nonblocking and rearrangeable asymmetric switching networks composed of digital symmetrical switches. *Unpublished manuscript*.

Jajszczyk, A. 1983. On nonblocking switching networks composed of digital symmetrical matrices. *IEEE Trans. Commun.*, **31**, 2–8.

Kabacinski, W. 1995. On nonblocking switching networks for multichannel connections. *IEEE Trans. Commun.*, **43**, 222–224.

Lin, G.-H., Du, D.-Z., Hu, X.-D., & Xue, G. 1999. On rearrangeability of multirate Clos networks. *SIAM J. Comput.*, **28**, 1225–1231.

Liptopoulos, F. K., & Chalasani, S. 1994. Nonblocking operation of asymmetrical Clos networks. *Pages 101–108 of: Proc. 23rd Int. Conf. Para. Proc.*, vol. 1.

Liptopoulos, F. K., & Chalasani, S. 1996. Semi-rearrangeably nonblocking operation of Clos networks in the multirate environment. *ACM/IEEE. Trans. Network.*, Vol.4, No.2, 281-291, Apr. 1996.

Little, C. H. C., Tutte, W. T., & Younger, D. H. 1988. A theorem on integer flow. *Ars. Combin.*, **26A**, 109–112.

Melen, R., & Turner, J. S. 1989. Nonblocking multirate networks. *SIAM J. Comput.*, **18**, 301–313.

Ngo, H. Q. 2003. A new routing algorithm for multirate rearrangeable Clos networks. *Theor. Comput. Sci.*, **290**, 2157–2167.

Ngo, H. Q., & Vu, V. H. 2003. Multirate rearrangeable Clos networks and a generalized edge coloring problem on bipartite graphs. SODA 2003: 834–840.

Niestegge, G. 1988. Nonblocking multirate switching networks. *In:* Bonatti, M., & Decina, M. (eds), *Traffic Engineering for ISDN Design and Planning.* Amsterdam: Elsevier.

Ohta, S. 1991. A simple control algorithm for rearrangeable switching networks with time division multiplexed links. *IEEE J. SAC*, **5**, 1302–1308.

Tsai, K. H., Wang, D. W., & Hwang, F. K. 2001. Lower bounds of wide-sense nonblocking Clos networks. *Theor. Comput. Sci.*, **261**, 323–328.

Varma, A., & Chalasani, S. 1993. Asymmetric multiconnection three-stage Clos networks. *Networks*, **23**, 427–439.

Index

Printed in the United States
By Bookmasters